Intelligence Without Brains

How Cells, Tissues, and Life Itself Solve Problems

Noble Trex

Published by NobleTrex Press

For permissions and other inquiries, write to:
P.O. Box 3132, Framingham, MA 01701,
USA

Contents

Introduction

Imagine a hunter stalking its prey through a dense, chaotic forest. The hunter moves with calculated silence, pausing to sniff the air, readjusting its path as the target zigs and zags. It anticipates the prey's desperation, cutting off escape routes with a burst of speed before finally engulfing the victim.

If I asked you to picture this scene, you might visualize a wolf in the Taiga or perhaps a cheetah on the Serengeti. You would naturally assume that this hunter possesses a brain-a central command center processing sights, smells, and strategies. You would assume it has eyes to see and muscles to fire.

But the hunter I am describing has none of these things. It is a neutrophil-a single white blood cell-chasing a distinct *Staphylococcus aureus* bacterium through the microscopic forest of blood cells on a glass slide. It has no eyes, yet it perceives. It has no muscles, yet it moves. Most importantly, it has no neurons, no cortex, and no brain. Yet, when you watch the footage of this chase, captured in grainy black and white over seventy years ago, you cannot escape the impression that this tiny blob of protoplasm is *thinking*. It makes errors, corrects them, pursues a goal, and ultimately succeeds. Modern intravital microscopy studies, such as Ronald Germain's group in 2013, make the same point in vivid detail.

For centuries, we have told ourselves a very specific story about who we are and where our minds come from. It is the story of the Neuron Doctrine. It says that intelligence is the exclusive province of the brain, that consciousness is a magic trick performed by the firing of billions of specialized nerve cells, and that the rest of the body is merely a mechanical carriage for the head. In this view, your liver, your skin, and your bones are dumb lumber-hardware that follows instructions but has no say in the matter.

This story is neat, tidy, and comfortably familiar. It is also wrong.

We are standing on the precipice of a revolution in the biological sciences, one that forces us to dismantle our neuro-centric view of the universe. The evidence is mounting that intelligence is not a luxury evolved by complex animals with skulls; it is a fundamental property of life itself. From the humblest bacteria to the cells organizing the architecture of your face right now, life is engaged in a constant process of problem-solving. It remembers, it plans, it negotiates, and it makes decisions.

To understand this is to realize that you are not a "ghost in a machine." You are a collective. You are a walking, breathing swarm of billions of tiny, intelligent agents, all working together in a precarious harmony to maintain the hallucination that you are a single individual.

The question that drives this investigation is not "do cells think?"- because, by any meaningful definition of problem-solving, they clearly do. The question is: *how?* How does a collection of mindless molecules create a mind? How does an embryo know how to build a heart on the left and a liver on the right? And what happens when we learn to speak the language of this cellular intelligence?

The Definition of the Spark

To see this hidden world, we must first abandon the baggage we carry about the word "intelligence." We tend to equate intelligence with human-level consciousness-Shakespeare, calculus, existential dread. But that is a penthouse view of a very tall skyscraper. Down in the basement, and on the ground floor, intelligence looks different. It looks like *competence*.

William James, the father of American psychology, offered a definition of intelligence that requires no neurons: the ability to reach the same goal by different means. If you are a wind-up toy and I kick you, you fall over and your gears spin helplessly. You have no agency. But if you are a mouse and I block your path to the cheese, you find another way. You adapt. You solve the problem of the maze.

What we are discovering, with increasing clarity, is that cells are more like mice than wind-up toys. They inhabit a world of chemical gradients, electrical fields, and mechanical pressures. They possess goals-maintain this temperature, build this organ, migrate to this location-and they possess the ingenuity to achieve those goals even when the rules of the game change.

This is not a metaphor. When we look closely at the machinery of life, we do not find a static blueprint. We find a feedback loop. We find a system that knows what it wants to look like and will work tirelessly to match its reality to that internal vision. This is the concept of "Unconventional Terrestrial Intelligence" (SUTI), and it suggests that cognition is not something that appears suddenly when you stack enough neurons together. It is a continuum that stretches all the way down to the bottom of the biological ladder.

The Electric Conversation

If cells are intelligent agents, how do they coordinate? How does a skin cell know it is part of a finger and not an eye? For decades, biology has been obsessed with genetics. We cracked the human genome and expected to find the instruction manual for the body. We thought that if we just read the DNA, we would know how to build a human.

But the genome is not a blueprint. It is a parts list. It tells you how to make the proteins, the molecular bricks and mortar. It does not tell you how to build the house. You can have all the correct bricks in a pile, but that doesn't make a cathedral. The missing piece of the puzzle-the architect that guides the construction-is bioelectricity.

Long before neurons evolved to carry high-speed text messages to the brain, humble cells were using ion channels to chat with their neighbors. Every cell in your body is a tiny battery, maintaining a voltage across its membrane. By tweaking these voltages, cells form electrical networks that store information, not about the present, but about the *future*. They store memories of what the body *should* look like.

In the coming chapters, we will walk through the laboratories where this electric code is being deciphered. We will meet the planarian flatworm, a creature that challenges every assumption we have about death and identity. You can chop a planarian into 279 pieces, and each piece will regenerate a perfect, miniature worm, as emphasized in Michael Levin's lab reviews in 2014. The question is: where is the memory of the worm stored? It's not in the brain, because you cut that off. It is stored in the bioelectric network of the tissue itself-a holographic memory of the self that persists even when the anatomy is destroyed.

We will see how researchers can hack this bioelectric software. By simply shifting the voltage of a few cells, we can induce a frog to grow an eye on its gut, as Pai and Levin showed in 2012, or convince a flatworm to regenerate a head at both ends of its body. These two-headed worms are not genetic mutants; their DNA is normal, as documented by Durant and Levin in 2017. They are creatures whose "target morphology"-their internal idea of what they should be-has been rewritten.

When the Collective Breaks

Understanding this cellular intelligence is not just an academic exercise. It is a matter of life and death. When the communication between cells breaks down, when the collective intelligence fractures, we get disease.

Consider cancer. We usually think of cancer as a genetic misfortune, a mutation that makes cells grow uncontrollably. But viewed through the lens of cognitive biology, cancer looks different. It looks like a reversion. A cancer cell is a cell that has stopped listening to the swarm. It has shrunk its "cognitive horizon." It no longer cares about the welfare of the organ; it cares only about its own replication. It is a single-celled organism trapped in a multicellular world, treating your body like the wild frontier.

This perspective offers a radical new hope. Instead of trying to kill cancer cells with toxic chemotherapy-a war of attrition that damages the patient-what if we could simply talk them back into the fold? What if we could re-establish the bioelectric connection, reminding the rogue cells of their duty to the collective? We will examine experiments that have done exactly this, such as Chernet and Levin's 2014 Xenopus work showing long-range bioelectric suppression of oncogene-driven growths.

Robots Made of Meat

The implications of taking cellular intelligence seriously extend far beyond medicine. They force us to rethink the nature of the artificial. For years, our best engineers have tried to build robots that can navigate the world, heal themselves, and adapt to the unforeseen. It is incredibly difficult. But nature solved this problem billions of years ago.

We are entering the era of the "Xenobot"-programmable living organisms built from the ground up using skin and heart cells, first demonstrated by Kriegman, Blackiston, Levin, and Bongard in 2020 and extended in 2021. These are not animals found in nature; they are engineered creatures, designed by supercomputers and built by biology. They walk, they swim, they work together, and they heal if you cut them. They blur the line between a born organism and a manufactured machine.

These living robots prove that the "self" is plastic. The genetic code of a frog cell usually builds a tadpole. But liberate that cell from the tyranny of the frog's anatomy, and it can build entirely new forms of life. The space of possible creatures is vast-a "morphospace" of alien shapes and functions that evolution has never explored, but which we are now beginning to map.

The Agency of Matter

As we travel through the landscape of the un-brained thinkers, from the slime molds that solve transit problems to the embryos that repair their own birth defects, a new picture of the universe emerges. It is a universe where matter is not passive. It is active, striving, and capable.

This shift in perspective is disorienting. It suggests that the "I" you feel behind your eyes is not a monarch ruling over a kingdom of stones, but a spokesperson for a vast, bustling democracy of micro-agents. It suggests that intelligence is a fundamental capacity of living matter, waiting to be unlocked.

But this view is also liberating. It means we are not alone in the dark. It means that the solutions to our most complex problems-how to regrow a severed limb, how to cure cancer, how to build resilient robotics-are already running on the software of the cells inside us. We just need to learn how to prompt them.

We are about to peer under the hood of life's engine. We will look at the machinery of desire and goal-seeking that hums within the cytoplasm. We will find that the ghost in the machine is real, but it isn't a spirit. It is a loop of electricity, a memory in the membrane, a decision made in the dark.

Welcome to the cognition of the wild.

Chapter 1

The Unseen Thinkers

We instinctively equate intelligence with the grey matter in our skulls, ignoring the billions of years life flourished before neurons existed. By reframing intelligence as the ability to solve problems in novel environments rather than the capacity to feel, we reveal a universe teeming with unconventional cognition hidden in plain sight.

1.1 The Brain Bias

A good way to start an argument about intelligence is to admit a prejudice. Ours is easy to spot: when we hear the word *thinking*, we picture a brain. Not just any brain, but usually a human one—wrinkled cortex, sparks of activity, perhaps a faint glow in a scanner image. The very grammar of the topic nudges us that way. We talk about *brainpower*. We call a clever person a *brain*. In everyday speech, intelligence is something you *have in your head*, as if the rest of the body were merely transport and plumbing.

This bias is so familiar that it hides in plain sight. A child can learn it without being taught. If you ask where your memories live, you point to your skull. If you ask what makes you *you*, you point there again. The story feels natural because it fits an undeniable fact: nervous systems can do astonishing things, fast. They can take streams of light on a retina, vibrations on an eardrum, chemical traces in a nostril, and turn them into a decision and an action in a fraction of a second. In an animal that must flee, chase, and mate, speed can be the difference between leaving offspring and leaving a corpse.

But speed, however impressive, is not the same as intelligence. It is one way of being competent in the world, not the definition of competence itself. The uncomfortable possibility raised by modern biology is that nervous systems did not invent the basic ingredients of information processing. They inherited them, tuned them, and concentrated them into specialized tissues. If that is true, then the question shifts: brains may be the most conspicuous example of biological intelligence, but not the only one, and not necessarily the original one.

The easiest way to see the brain bias at work is to notice how quickly we become suspicious when a system behaves intelligently without a brain. A plant leans toward light and learns the rhythm of days; we call it *tropism* or *entrainment*, not learning. A bacterium climbs a chemical gradient; we call it *chemotaxis*, not seeking. An embryo repairs itself after damage; we call it *developmental regulation*, not problem-solving. The vocabulary is not merely cosmetic. Words determine what kinds of explanations we allow. If we reserve terms like *decision*, *memory*, and *goal* for brains, then non-brain systems are, by definition, never doing those things. They are merely executing chemistry.

This linguistic gatekeeping has a history, and it is not a trivial one. In the late nineteenth and early twentieth centuries, neuroscience coalesced around a powerful organizing idea: the nervous system is made of discrete cells called neurons, and the neuron is the fundamental unit of computation. This idea is so successful that it has become invisible. It shaped how laboratories were built, what instruments were developed, what counted as a signal, and even what questions were considered respectable. It is difficult to overstate the influence of Santiago Ramón y Cajal, whose meticulous drawings and arguments helped establish what became known as the neuron doctrine: the nervous system is not a continuous web but a network of individual cells that communicate at specialized junctions.

Cajal did not merely describe neurons; he offered a worldview. If the nervous system is a network of specialized excitable cells, then the natural home of information processing is a dedicated tissue. From there it is a short step, culturally and scientifically, to the assumption that intelligence *requires* such a tissue. This assumption becomes particularly tempting in humans because so much of what we value—language, abstract reasoning, planning across decades—is closely tied to cortical activity. When we damage certain brain regions, capacities vanish. When we stimulate others, perceptions appear. When we record electrical patterns in motor cortex, we can decode intended movement. The evidential weight is real.

Yet the neuron doctrine, like many successful ideas, can be overextended. It can become a doctrine not only about what neurons are, but about what thinking must be. That is where the trouble begins.

A useful corrective comes from a simple question: what exactly is a neuron that makes it so special? Strip away the romance and the metaphors. A neuron is a cell with a membrane. That membrane contains proteins—ion channels and pumps—that control the flow of charged particles such as sodium, potassium, calcium, and chloride. The controlled flow of ions creates voltage differences across the membrane, a kind of electrical tension. Changes in that voltage can propagate, triggering the release of chemical messengers or the flow of current through junctions to other cells. In other words, a neuron is a cell that uses bioelectricity to sense, integrate, and act.

Now notice something that should feel slightly unsettling: neurons are not the only cells with membranes, ion channels, pumps, and voltage differences. Every living cell maintains an electrical state. The details differ—the specific channels, the time scales, the coupling to chemical pathways—but the basic hardware is ancient. It is older than animals. It is older than nervous systems. It is one of life's fundamental tricks: use electrochemical gradients to organize matter and to coordinate actions.

If that sounds abstract, it helps to look at a diagram that refuses to flatter our intuitions.

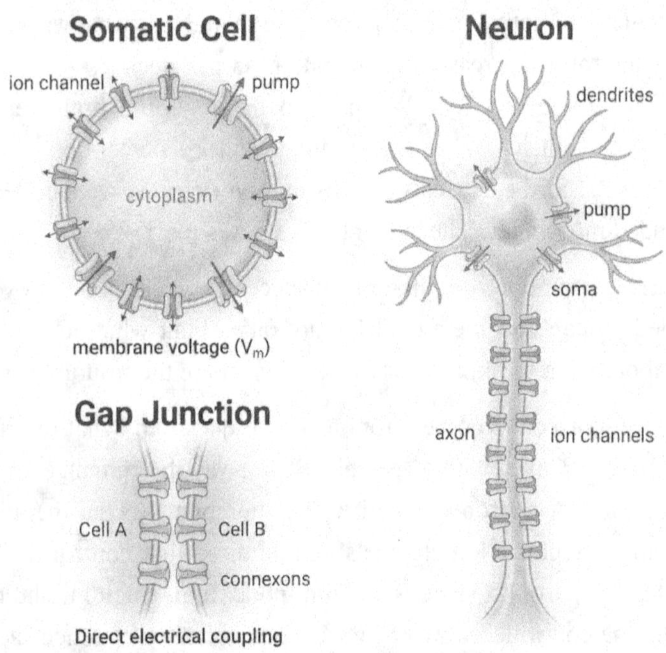

Evolutionary continuity. Somatic cells (top) maintain membrane voltage (V_m) using ion channels and pumps. Neurons (middle) specialize this machinery for rapid long-distance signaling via axons. Gap junctions (bottom) allow direct cytoplasmic and electrical coupling between cells, forming networks independent of synapses.

Once you have this picture in mind, it becomes harder to maintain the comforting belief that intelligence begins when evolution invents neurons. Neurons are an elaboration, not an ex nihilo creation.

They are a way of taking a general cellular capacity—bioelectrical control—and packaging it for speed and long-distance communication in multicellular bodies.

Why, then, do we keep returning to brains as the exclusive seat of intelligence? Partly because brains are spectacular. They produce behaviour that maps neatly onto our own introspection: we feel ourselves perceiving, deciding, remembering. It is tempting to treat that feeling as a guide to mechanism. But introspection is a poor microscope. It tells us what it is like to be a system with a certain architecture; it does not tell us which parts of that architecture are necessary for competence.

There is also a sociological reason. Neuroscience became a prestige science in the twentieth century. Its technologies were photogenic: electrodes, oscilloscope traces, coloured brain scans. Its metaphors fit the era: circuits, wiring diagrams, information processing. It offered a clean division of labour: cognition in the nervous system, support in the rest. That division does not merely simplify textbooks; it simplifies funding, departmental boundaries, and intellectual identity. If a phenomenon can be framed as neural, it joins a well-developed research tradition. If it cannot, it risks being dismissed as metaphor or mere physiology.

The plant neurobiology controversy, which flared in the early 2000s, illustrates how deeply this bias runs. Some researchers proposed that plants, though lacking neurons, exhibit signalling and integrative behaviour significant enough to justify a field named by analogy: *plant neurobiology*. Critics responded sharply that the term was misleading, even sensational, because plants have no neurons and no brain. Supporters replied that the point was not to claim tiny green brains but to draw attention to genuine information processing in plants: electrical signals, chemical communication, memory-like effects, coordinated responses to stress. Beneath the argument about terminology was a more consequential disagreement about what counts as cognition. Does cognition require neurons, or does it require something more abstract: the ability to sense, integrate, and act adaptively across time?

It is worth pausing over the emotional temperature of such debates. The phrase "plants think" triggers discomfort in many people, including scientists who otherwise accept that plants are dynamic, responsive, and complex. Why discomfort? Because it threatens a cherished boundary. Humans have long used cognitive capacity as a moral and social dividing line: between people and animals, between animals and plants, between the living and the inert. If plants have even a faint cousin of what we call mind, the line blurs. That blurring can feel like a loss of status, or like an invitation to anthropomorphism. The safest move, psychologically, is to insist that without neurons there is no cognition. The brain bias is not only a scientific habit; it is a cultural comfort.

Modern biology presses against that comfort from multiple directions. One direction is developmental biology: the study of how an embryo builds a body. Consider what an embryo must do. It begins as a single cell, then becomes millions. Those cells must differentiate into dozens of types, migrate, fold into organs, wire themselves, and stop at the right size and shape. If you perturb the process—remove cells, cut tissues, change chemical gradients—development often corrects itself. The system is not a simple chain reaction. It is robust and goal-directed in a practical sense: it tends to reach a reliable anatomical outcome despite disturbances.

Another direction is regeneration. Some animals can rebuild complex structures after injury: limbs, tails, even parts of the brain. Regeneration is not mere growth. It requires knowing what is missing, where the boundary of the body should be, and when to stop. In human terms we might call this *pattern memory*: a stored specification of a desired form. The mystery is where that specification lives and how it is enforced. Neurons can influence regeneration, but regeneration occurs in organisms and contexts where neural input is absent or minimal. The control is distributed across tissues.

Bioelectricity provides a bridge between these observations and a deeper rethinking of intelligence. Cells in tissues are electrically active not only in the fleeting spikes of neurons but in slower, spatially extended patterns of membrane voltage. Cells can be coupled by gap junctions, forming networks in which voltage states propagate and influence gene expression, cell movement, and growth. Such networks can store information in stable patterns, switch between states, and coordinate action across a tissue. If that sounds like computation, it is because computation is not a substance; it is an organized relationship between states, signals, and outcomes.

This is one reason the old habit of equating cognition with neuron count is increasingly fragile. We can respect the extraordinary power of nervous systems without assuming that neurons are the only way to build a problem-solving controller. Evolution is opportunistic. It reuses parts. It builds new layers on old platforms. The bioelectrical machinery that makes neurons possible is also available to every other cell. The question is not whether non-neural tissues can ever look like a brain, but whether they can achieve some of the same *functions*: integrating information, maintaining internal goals, correcting errors, and coordinating action across a body.

The bias toward brains also obscures a subtle point about time. Neural cognition is fast, and speed is visible. But many of life's most important problems are slow. Building a body takes days or months. Repairing tissue takes weeks. Maintaining stable physiology takes a lifetime. These are not tasks that benefit from millisecond spikes alone. They require sustained, robust control. A system can be intelligent on a time scale that does not resemble a thought in our head.

Once you grant that possibility, another prejudice begins to loosen: the belief that intelligence must be accompanied by a rich inner movie. People often assume that if a system is intelligent it must, somewhere inside, *feel* like something. Yet biology routinely produces systems that are competent without being anything like a human subject. The temptation to smuggle in consciousness as a requirement is understandable—consciousness is the part we know from the inside—but it is also a trap. If we make subjective experience the yardstick, we will never develop a comparative science of intelligence across the tree of life. We will be left arguing about which organisms *really* have minds, rather than measuring what organisms can do and how they do it.

The brain bias, then, is not merely an error; it is a narrowing of imagination. It makes us overlook the ways in which tissues and cells already solve problems. It also makes us underestimate evolution's thrift: rather than inventing intelligence once, in nervous systems, evolution may have been refining and reconfiguring a more general capacity present in living matter from the beginning.

There is a practical consequence to getting this right. If intelligence is a set of capabilities rather than a particular organ, then it becomes thinkable to engineer or guide those capabilities in new contexts. Regenerative medicine, for instance, can be reframed as a negotiation with a tissue's own problem-solving machinery: persuading it to rebuild what we want, not merely forcing it to proliferate. Cancer can be viewed not only as runaway growth but as a breakdown in the collective coordination that normally keeps cells aligned with the body's larger goals. Even synthetic biology looks different if we treat cell collectives as agents navigating constraints rather than as passive chemical reactors.

But before we rush toward applications, we need a clean conceptual move. We must separate two things that our brain-centric intuitions fuse together: *intelligence* and *understanding*. A system can reach a goal reliably and yet have no idea, in any human sense, what it is doing. That unsettling separation is the next step, and it will take us from the romance of brains to a colder, more useful notion: competence without comprehension.

1.2 Competence Without Comprehension

Once you stop treating the brain as the sole passport to intelligence, an awkward question arrives almost immediately: what do we mean by *intelligence* in the first place? The temptation is to reach for the richest thing we know from the inside—conscious experience. If intelligence means having thoughts, then intelligence seems inseparable from a someone who is thinking them. But that move puts us back where we started, because it quietly smuggles in the very architecture we are trying not to privilege. Consciousness is the signature experience of brains like ours; it is not a neutral measuring stick.

A more useful definition has to be something you can apply to bacteria, embryos, and slime molds without turning the discussion into a seance. The definition this book will use is deliberately plain: *intelligence is the ability of a system to reach goals in the face of perturbations.* A goal, in this operational sense, is not a daydream. It is a stable end-state the system tends to return to or maintain—a body temperature, a target shape, a chemical balance, a safe location, a repaired wound. Perturbations are what the world does to you: noise, injury, missing information, changing conditions. An intelligent system is one that does not merely react; it *adjusts* so that it still gets where it needs to be.

This definition will feel too thin to some readers. Where are the beliefs and desires? Where is the narrative self, the inner voice, the feeling of deliberation? The answer is that those may be magnificent additions, but they are not the minimum for problem-solving. They are one way a system can control outcomes, not the only way. In fact, one of the most productive ideas in contemporary philosophy of mind is that competence can exist without comprehension: a system can behave in ways that look clever while lacking any explicit grasp of what it is doing or why.

If that phrase rings faintly familiar, it is because it has wandered into public conversation through the work of Daniel Dennett, who used it to emphasize an uncomfortable truth. Much of what appears purposeful in nature is the result of processes that do not themselves understand purposes. Natural selection produces exquisitely competent organisms without *anyone* in the loop planning the design. Even within an organism, many subsystems behave as if they had goals—maintain blood sugar, keep pH within bounds, repair DNA damage—without needing a tiny homunculus reading dashboards. Competence comes first; comprehension, when it appears, is a special case.

The phrase is not merely a philosophical provocation. It gives biologists a licence to ask a pragmatic question: *what does the system reliably achieve, and by what mechanisms?* That question is gentle enough to be asked of a bacterium and strict enough to keep us honest. It also forces a discipline that matters for this book: to be careful with metaphors. Saying that a cell "wants" to heal can be a useful shorthand, but we should always be able to translate the shorthand into a description of feedback, signaling, and control.

A historical anecdote helps here, because our species has faced a similar conceptual shift before. In the early age of steam engines and mechanical regulators, engineers learned to build devices that seemed uncannily purposeful. The classic example is James Watt's centrifugal governor, a spinning mechanism with two metal balls that rise as the engine speeds up and fall as it slows. Through a linkage, the balls adjust a valve that controls steam input. If the engine runs too fast, the balls fly outward and upward, the valve closes a bit, and the engine slows. If the engine slows, the balls fall, the valve opens, and the engine speeds up. The governor keeps the engine near a stable speed across changes in load, without a human hand continuously tweaking it.

People did not accuse Watt's governor of having consciousness. Yet it does something recognizably intelligent by our operational definition: it reaches and maintains a goal state (a steady speed) despite perturbations (changing loads, fuel variability). It senses an error, integrates it into a corrective action, and stabilizes the system. This is the heart of feedback control.

Living systems are stuffed with governors. Some are chemical, some electrical, some mechanical, and many are hybrids. What makes life eerie, compared with machines, is the range of goals it can maintain and the creativity of its methods. A bacterium does not merely hold a set point; it can climb a gradient. A tissue does not merely remain intact; it can rebuild a missing part. The underlying logic, however, is often familiar to an engineer: feedback, memory, and adaptive tuning.

Bacterial chemotaxis—a bacterium's ability to move toward attractants like nutrients and away from repellents—is a miniature masterpiece of competence without comprehension. The organism is microscopic; it has no brain, no neurons, no centralized controller. It is, in a sense, all sensor and all motor at once. Yet it can find food in a world where molecules diffuse and currents swirl, where information is noisy and delayed. How?

A common bacterium such as *E. coli* swims by rotating flagella, tail-like filaments that work like propellers. When the flagella rotate together, the bacterium moves in a roughly straight 'run'. When rotation changes, the bacterium 'tumbles,' reorienting randomly. By modulating the probability of tumbling, the bacterium biases its random walk. If conditions are improving, it tumbles less and continues running; if conditions worsen, it tumbles more, trying a new direction. Over time, this simple strategy produces a drift toward higher concentrations of attractant.

This already looks clever, but the deeper intelligence is in how the bacterium handles scale. If the definition of "good" were fixed, the system would fail in real environments. A concentration that is high in one context might be low in another. Sensors saturate. Noise dominates at low signal levels. The bacterium needs to stay sensitive across orders of magnitude, and it needs to avoid getting stuck reacting forever to a constant stimulus. If a system simply "turns on" when it detects food and stays on, it loses the ability to detect changes. The world becomes a blinding light.

The elegant solution is *adaptation*: the ability to respond to a change and then reset sensitivity even while the new condition persists. In chemotaxis, bacteria exhibit something close to "perfect adaptation": after an initial response to a step change in attractant, the internal signaling output returns near its baseline, ready to detect further changes. Engineers recognize this as a hallmark of a particular design principle: integral feedback control, in which the system effectively accumulates error over time and uses that integral to drive correction. The language sounds technical, but the intuition is domestic. If your shower water is too cold, you don't merely twist the knob a fixed amount and accept whatever happens. You keep adjusting until the temperature matches your preference, and you can do that reliably even if water pressure changes. The mechanism that makes such robustness possible is a form of integral action.

Two points matter for our purposes. First, the bacterium achieves a goal (climbing a gradient) not by "understanding" gradients but by embodying a control strategy in molecular networks. Second, this competence depends on the structure of feedback. It is not magic; it is an architecture.

At this point, a skeptic might concede that bacteria are clever but insist that this is still not *intelligence* in any meaningful sense. Isn't it just chemistry? The answer is yes—and that is precisely the point. If intelligence is a way of describing successful control in the face of uncertainty, then chemistry can be intelligent when arranged in the right loops. The cognitive vocabulary is not an insult to mechanism; it is a description of what the mechanism accomplishes.

Chemotaxis offers an additional lesson: competence often costs energy. In everyday life, we associate intelligence with energy consumption—the brain is metabolically expensive—but we usually treat sensing as passive, like a thermometer reading the air. In biology, many sensing strategies are *active*. They consume energy to maintain sensitivity, reduce noise, and avoid being trapped by equilibrium limitations. Some chemotaxis models suggest that to reproduce certain observed features of bacterial response, you need nonequilibrium cycles powered by ATP, the cell's energy currency. This is not a pedantic biochemical detail; it is a conceptual point. Achieving reliable goal-directed behavior in a noisy world can require paying a thermodynamic price. Competence is often powered.

The same theme appears at larger scales in organisms that look, at first glance, like they are doing something even more mind-like. Slime molds are a classic example, not because they are mystical but because they sit in a sweet spot: large enough to watch with the naked eye, simple enough to embarrass our categories. *Physarum polycephalum*, the famous yellow plasmodial slime mold, is a single giant cell with many nuclei. It can spread across a surface as a veined network, exploring for food. It can form transport-efficient structures, pruning and reinforcing tubes in ways that resemble a crude form of planning. It can solve simple mazes in laboratory demonstrations, not by thinking about mazes but by growing, retracting, and redistributing flow.

For a long time, these feats invited anthropomorphic commentary: the slime mold "chooses" the shortest path, "decides" between options, "remembers" where it has been. Those words can be useful if they remain placeholders for mechanisms. What mechanisms might ground them? One increasingly concrete account focuses on the organism's internal dynamics: rhythmic contraction waves that drive fluid flow through its network. The slime mold is, in effect, a living pump system. Waves of contraction move cytoplasm back and forth, and the resulting flows redistribute nutrients and signaling molecules. Under constraints—a bright region the slime mold tends to avoid, a gap it must cross, a boundary it must circumvent—the contraction patterns can shift. Over time, the organism can settle into modes of contraction and flow that are more efficient for escape or transport. The "decision" is not a verdict delivered by a central judge; it is an emergent stabilization of a dynamical pattern that achieves the goal.

This is competence without comprehension in a visceral form. You can watch the slime mold hesitate at a boundary, extend protrusions, retreat, try again, and eventually commit. The behavior has the texture of deliberation, yet the mechanism may be a distributed negotiation among local contractions, mechanical forces, and feedback from the environment. The organism does not need to represent the problem in symbols. It needs to explore, sense error, and settle into a pattern that works.

We are now close to the operational definition's real payoff: it allows comparisons across scales and substrates. A bacterium's molecular circuit and a slime mold's peristaltic network look nothing alike, but both can be understood as controllers steering a system through disturbances toward a stable outcome. Likewise, when we consider tissues in a developing embryo, we can ask: what are the goals, what are the error signals, what are the feedback loops?

This framing also helps with a common confusion about purpose. When biologists say a system is goal-directed, they do not need to claim that the system has foresight. Goal-directedness can be entirely retrospective, entirely mechanistic. A ball rolling down a hill is not goal-directed in this sense; it is merely following physics. But a system that corrects deviations—that behaves differently depending on how far it is from a preferred state—is doing something more interesting. The existence of error correction is the tell. The system acts as if it has a destination, because its dynamics are organized around reaching and maintaining one.

A worry remains: if we define intelligence as goal-seeking under perturbation, do we risk calling *everything* intelligent? A thermostat, a soap bubble, a river carving a channel? The boundary is real, and it matters. A soap bubble returns to a sphere after deformation because surface tension minimizes area; there is no sensing, no memory, no ability to alter strategy. A thermostat has a simple sensor and actuator but a narrow repertoire and limited capacity to cope with new contexts. Biological systems, by contrast, typically show a richer kind of control: multiple goals held in tension, adaptive sensitivity across changing conditions, and the capacity to reorganize their own internal wiring. The definition does not flatten distinctions; it invites us to measure them.

That measurement can be practical. If intelligence is about reaching goals, we can ask how many different goals a system can reliably achieve, across what range of conditions, how quickly it recovers from disturbance, and whether it can generalize from past perturbations. These are not mystical questions. They can be turned into experiments. They can be quantified. They also respect the diversity of life's architectures.

This is where the previous section's discussion of ancient bioelectric hardware quietly returns. If cells and tissues have electrical states, and if those states can be coupled across networks, then tissues may possess their own forms of memory and error correction. A tissue's "goal" might be an anatomical pattern, and its "perturbation" might be an injury. The mechanisms could involve gene regulation, mechanical forces, and bioelectrical coupling all at once. The point is not to pretend a tissue is consciously planning. The point is to recognize a kind of competence that is easy to miss when we look only for neurons.

The operational definition also clarifies why this book will mostly sidestep the question of sentience. Sentience matters ethically; it matters personally; it matters politically. But it is not necessary for the scientific project at hand. You can build a rigorous theory of flight without first solving the philosophy of "what it feels like to be airborne." Likewise, we can study intelligence as control and problem-solving without settling whether a slime mold has experiences. That restraint is not evasive; it is what allows a comparative science to exist at all.

If there is an emotional sting in this approach, it is because it demotes a cherished human intuition: that understanding is the essence of intelligence. Much of our pride lies in comprehension, in being able to say why we chose what we chose. Yet many of our own competencies are not like that. We can catch a falling glass without computing trajectories consciously. We can speak grammatically without knowing the rules we follow. We can recognize a face instantly without being able to list the features that drove the recognition. Human minds themselves are layered: pockets of comprehension floating on oceans of competence.

Seen that way, the book's definition is not an insult to intelligence; it is a more general portrait of it. Intelligence becomes something life does, at many levels, whenever it builds systems that can steer themselves toward stable outcomes in a shifting world. Brains are one spectacular expression of this, but they are not the only stage on which competence appears.

The next move is to ask where, exactly, these goals and perturbations live. A bacterium's problem is not laid out in Euclidean space the way a maze is drawn on paper. An embryo's problem is not simply "grow" but "grow into *this* form." A tissue's problem may be distributed across chemical gradients, electrical states, and mechanical stresses. To understand intelligence without brains, we need a map—not of geography in the ordinary sense, but of the abstract landscapes of possible states in which living systems act. That landscape, and how organisms navigate it, is where we turn next.

1.3 The Geography of Problems

If intelligence is the reliable reaching of goals under disturbance, then the next question is almost embarrassingly simple: where, exactly, are those goals? Not in the sense of where they are *located* in an organism, but where they live as targets in the space of what is possible. A thermostat's goal can be written as a number. A bacterium's goal can be drawn as an arrow up a chemical gradient. A tissue's goal might be an anatomical pattern—a limb with the right length and symmetry—that exists nowhere in the outside world until the tissue builds it.

To talk about such goals without immediately falling back into brain metaphors, it helps to borrow a geometric idea. Many problems are not best pictured in ordinary three-dimensional space. They are better pictured in a *problem space*: an abstract landscape in which each point represents a possible state of a system, and movement across the landscape represents change over time. Some regions are good (stable, viable, safe, functional). Others are bad (unstable, damaging, dead ends). In this picture, intelligence becomes a kind of navigation.

At first this seems like a metaphor, the kind you might indulge at a dinner party and later regret in the lab. But the reason it is worth taking seriously is that it compresses a great deal of messy biology into a form we can reason with. It does not tell us what the mechanisms are, but it tells us what those mechanisms must *achieve*: they must guide the system away from traps and toward basins of stability, even when the wind changes.

A basin of stability is a useful notion here. Imagine a system that, if nudged, tends to return to a particular state. A marble in a bowl will roll back to the bottom after you tap the bowl. The bottom is a stable point. In biology, homeostasis—the maintenance of internal conditions within life-supporting ranges—is a web of such basins. Your blood pH, your body temperature, your blood glucose levels: each has dynamics that resist drift. The remarkable feature is not that organisms have stable states; many physical systems do. It is that organisms can maintain stability across a wide range of conditions, and can sometimes shift deliberately between stable states when circumstances demand it.

To make the idea concrete, it helps to visualize a problem space as a topographic landscape. Valleys are traps or stable basins; ridges are barriers; plateaus are regions of weak guidance; and the "agent" is not a heroic explorer but a system with limited information and limited control. The path it takes is a history of decisions, corrections, and compromises.

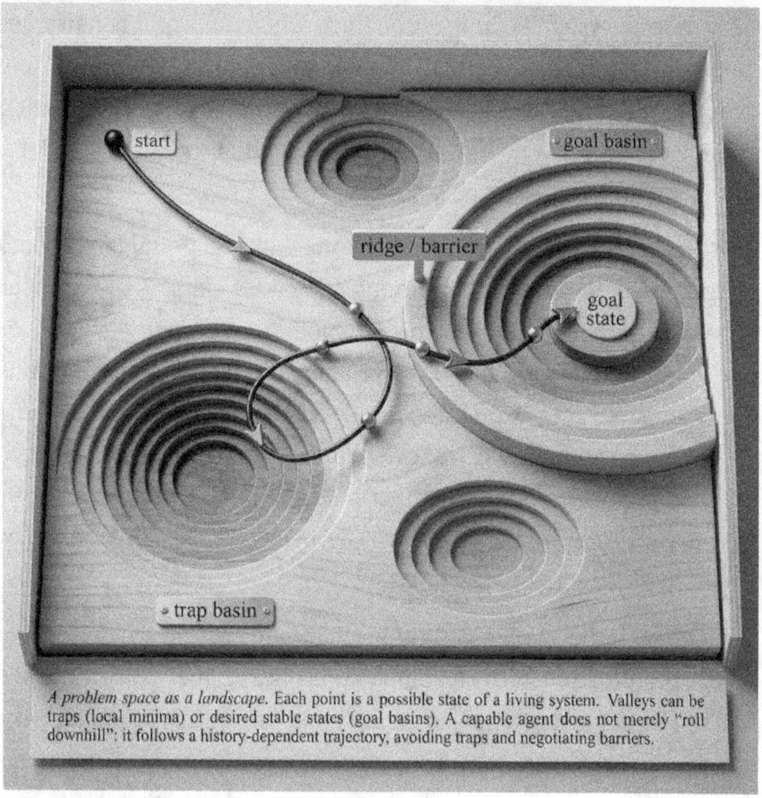

A problem space as a landscape. Each point is a possible state of a living system. Valleys can be traps (local minima) or desired stable states (goal basins). A capable agent does not merely "roll downhill": it follows a history-dependent trajectory, avoiding traps and negotiating barriers.

This picture does something subtle. It loosens the grip of a familiar story we tell about agency: that it must be a little captain inside the body, looking at the world and issuing commands. In the landscape view, agency is not a ghost in a cockpit. It is a capacity of the whole system to move through its state space in a way that keeps it viable and, sometimes, improves its prospects.

That definition of agency is deliberately cybernetic. Cybernetics, a mid-twentieth-century movement that studied control and communication in animals and machines, insisted that goal-directed behavior could be understood through feedback loops rather than inner narratives. The language can sound cold, but it was motivated by wonder.

How does a system keep itself on track when the world is noisy and unpredictable? The cybernetic answer was: it must sense deviations from a preferred state and act to reduce them. Whether the system is a pilot correcting a plane's heading, or a cell correcting its internal chemistry, the logic can be similar.

But the landscape metaphor adds something that simple feedback diagrams often miss: not all problems have a single knob you can turn. Many biological problems are high-dimensional. "Where you are" in problem space may involve temperature, ion concentrations, tissue tension, gene expression levels, and the presence of neighbors. The same perturbation can shove you in multiple directions at once. A good controller is not one that reacts strongly; it is one that reacts *appropriately*, avoiding responses that solve one subproblem by creating a worse one.

This is why the earlier definition of intelligence as goal-reaching under perturbation needs a companion concept: the *geometry* of the goal. The geometry tells you what counts as an improvement, what counts as a trap, and how costly various routes are. Two organisms can face the "same" physical environment and yet inhabit different problem spaces, because they have different sensors, different effectors, and different internal constraints. A puddle is a trivial obstacle to a bird and a deadly ocean to an ant. The physics is identical; the problem is not.

The biologist Jakob von Uexküll made this point vivid more than a century ago with a word that deserves to be better known: *Umwelt*. It means, roughly, the organism's self-centered world—the slice of reality that matters to it because it can sense it and act on it. Uexküll's most famous character is the tick. A tick perched on a branch does not survey the forest like a miniature mammal. It does not map the terrain, weigh options, and daydream about the future. Its Umwelt, as Uexküll described it, is built around a few cues: the smell of butyric acid (a component of mammalian sweat), the warmth of skin, the feel of fur. Those cues are enough. When the right chemical wafts up, the tick lets go. When it lands on warm skin, it crawls. When it feels fur, it burrows. From a human viewpoint this is impoverished perception. From the tick's viewpoint it is a complete world, because it contains what the tick needs to solve its problems.

The tick anecdote is sometimes retold as a curiosity about simplicity. Its deeper lesson is about the geography of problems. A "problem" is not just out there in the environment; it is defined by the coupling between an organism's sensors and its effectors. Sensors carve reality into a few meaningful dimensions. Effectors define what changes the organism can actually produce. The resulting loop—sense, act, sense again—is the terrain on which agency happens. When we say an organism navigates a problem space, we mean it moves through the space of states that its own body and behavior make available.

This is why it is misleading to imagine a bacterium "climbing a gradient" as if it had a tiny map of chemical concentrations. What it has is a sensor that detects changes over time, a motor that can bias movement, and a feedback loop that couples the two. The "gradient" exists for the bacterium as a stream of better-or-worse signals along its path. The bacterium's problem space is not the lab technician's concentration plot; it is the bacterium's internal state as it experiences the world through its own loop.

The same point becomes more dramatic in multicellular life, where problems exist simultaneously in multiple spaces. Consider a developing embryo. The embryo's anatomical problem space is the set of possible body plans it might become. Its physiological problem space includes things like ion balance and metabolism. Its transcriptional problem space—the space of gene expression states—determines what kinds of cells exist and what they are capable of doing. These spaces are coupled. A change in membrane voltage can alter gene expression; a change in tissue mechanics can alter voltage; a change in gene expression can change the set of ion channels and thus reshape voltage dynamics. A perturbation in one space can push the system into a different region in another.

For anyone who has watched a cut heal, this coupling has a human immediacy. Your skin does not merely seal a gap; it coordinates inflammation, clotting, cell migration, proliferation, and remodeling, often restoring a functional barrier. A naive physical system would respond to a cut by continuing to leak. A naive chemical system might react violently and then exhaust itself. Healing is a controlled traversal: away from damage, through a sequence of transient states, toward stability. It is a path through problem space.

The idea that organisms inhabit problem spaces also corrects a common misunderstanding about "fitness" and "adaptation." We often imagine evolution as if it were shaping organisms to fit a fixed environment. But organisms do not merely adapt to environments; they actively reshape them. This is the core of niche construction theory: the idea that organisms modify the selection pressures they and their descendants experience by changing the environment itself. Earthworms alter soil structure. Beavers build dams. Bacteria alter pH and oxygen levels around them. Humans, on a scale that now startles even us, remodel landscapes and atmospheres.

From the landscape point of view, niche construction means the terrain is not fixed. The agent is, to some degree, moving the hills and valleys. This matters for our definition of agency. An agent is not just a system that moves within a given problem space; it is often a system that changes the shape of the space it must navigate. Some of this change is obvious, like a beaver dam. Some of it is subtle, like a microbial community that produces a chemical environment favoring its own members. Even within a body, cells modify their local environment by secreting extracellular matrix, changing stiffness, altering electric coupling through gap junctions, and adjusting the chemical milieu. Problems are not simply solved; they are sometimes *redefined.*

This brings us to a further complication that is, in fact, a gift: intelligence can be redistributed into the body itself. In robotics and biomechanics, this is sometimes called embodied intelligence or morphological computation: the idea that physical form can perform part of the "computation" that we might otherwise attribute to a central controller. A simple example is passive dynamic walking. With the right leg geometry and joint friction, a mechanical walker can take stable steps down a slight incline without motors or a computer calculating trajectories. Its body shape and gravity do much of the work.

Biology is full of such redistributions. The octopus's arms contain much of their own control; the brain issues broad commands, but the arm's mechanics and local circuits handle details. The human hand's tendons, elasticity, and friction allow fast grasp adjustments that would be hard to compute explicitly. Even at the cellular level, tissue architecture can make certain coordinated behaviors easier: aligned fibers guide cell migration; networks of gap junctions allow electrical states to spread; mechanical tension can propagate information about shape. The "mind," in the minimal sense of goal-directed control, need not sit in a single place.

Where does this leave agency? Not as a mystical property, and not as a synonym for consciousness, but as a measurable capacity: the extent to which a system can steer itself through its relevant problem spaces, keeping itself away from failure modes and moving toward stable, functional states. Agency becomes graded. A rock has almost none. A thermostat has a sliver. A bacterium has more. A tissue has more still, because it can coordinate across many parts and often correct structural errors. A brain-equipped animal may have agency in a richer set of spaces, including social ones.

This graded view avoids a common trap. If agency is an all-or-nothing property bestowed by a nervous system, then everything without neurons is passive. But if agency is the ability to keep certain variables within bounds and to reach certain targets despite disturbances, then agency appears wherever there is robust control. The question becomes: *which variables can the system control, and across what range?* A bacterium can control its position relative to chemical cues. A tissue can control its collective pattern. An organism can control aspects of its niche.

The landscape metaphor also helps clarify why the same outward behavior can reflect very different internal "intelligences." Two systems might both end up in the goal basin, but one might do so only under a narrow range of conditions, while the other can handle surprises. In problem space terms, one path is fragile; the other is robust. Robustness is not a decorative feature. In living systems, robustness is often the whole point, because life is what persists in the face of a world that does not cooperate.

If this sounds like it is drifting toward abstraction, it is worth returning to a concrete image: a wound that closes, an embryo that normalizes after being gently perturbed, a slime mold that reorganizes its flows to escape an unpleasant constraint. In each case, something is "steering." Yet there may be no central steering wheel. Instead, there are local sensors, local actions, and coupling that allows the whole to behave as if it had a goal.

The Search for Unconventional Terrestrial Intelligence (SUTI) is, at heart, a commitment to looking for that steering wherever it appears, without demanding that it look like our own. The method is not to romanticize microbes and tissues, but to identify the spaces they navigate and the variables they control. Once you begin to ask those questions, a surprising continuity emerges. The same control logic that stabilizes a bacterium's sensing can also stabilize a tissue's patterning. The same notion of a trap in a landscape can describe a local minimum in a slime mold's exploration and a pathological stable state in a diseased organ. The geography changes, but the idea of navigation persists.

This section has leaned heavily on metaphors of hills and valleys, but the reason is not poetic indulgence. It is preparation. In the chapters that follow, we will need a way to talk about tissues and cell collectives as agents without pretending they are tiny persons. The landscape idea supplies that: it lets us speak about goals, traps, and trajectories in a disciplined way, while leaving open the mechanistic details.

Those details are where things become truly strange, because biology does not navigate only in chemical concentrations and mechanical forces. It navigates in electrical patterns, too—slow, distributed voltage states spread across tissues, coupled by ion channels and gap junctions. If we want to understand how a body plan can be remembered, corrected, and rebuilt, we will need to examine how these bioelectric networks carve their own problem spaces, and how a living system might store a map of "where it ought to be" in patterns that look nothing like neurons firing.

Chapter 2

The Ghost is a Loop

Living systems often appear to be haunted by a guiding spirit or a blueprint, yet the driver of biological complexity is purely mechanical. By tracing the history of cybernetics and the machinery of feedback, this chapter reveals how 'purpose' emerges from the simple mathematics of error correction, transforming the ghost in the machine into a loop in the wiring.

2.1 The Governor and the Glitch

A steam engine is a straightforward kind of honesty. Feed it heat, give it water, let the pressure push a piston, and you get motion. In the early days of industrial power, that seemed like the whole story: a chain of causes marching forward. Coal becomes fire, fire becomes steam, steam becomes force. If the engine ran too fast or too slow, it was because something *upstream* had changed: the fire was stoked, the load was heavier, the valve was sticky. Add a clever operator and the machine could be kept in line.

Then factories began to demand something that did not fit this one-way picture. Not just power, but *steady* power. A loom does not appreciate surprise; neither does a milling machine. Too slow and the work backs up; too fast and parts break, threads snap, and people get hurt. The new industrial world wanted machines that behaved as if they had a preference: *keep the speed about here*. That sounds like purpose. It sounds like a little ghost sitting inside the ironwork, watching the dial and deciding when to ease off.

The striking thing is that engineers learned how to build that kind of apparent purpose without adding a mind, a map, or a manager. They did it by building a loop.

The loop has a particular feel to it when you first meet it. A familiar linear explanation says: *the throttle controls the engine speed.* A looped explanation says: the throttle controls the speed, but the speed also controls the throttle. Cause and effect chase each other in a circle, and the circle itself becomes the explanation. To people trained on straight arrows, this can sound like cheating: how can something be both cause and consequence? Yet this is exactly the logic that makes goal-directed behavior possible in machinery and, as we will see, in biology. The goal does not need to be written anywhere in words. It can be embodied in the way a system continually compares what is happening to what *should* be happening, and then acts to close the gap.

The classic emblem of this shift is the centrifugal governor, most famously associated with James Watt's improvements to steam engines in the late eighteenth century. It is an object that looks almost comically alive: two heavy balls on jointed arms, spinning around a vertical shaft like a pair of dancers holding hands. When the engine speeds up, the balls swing outward and upward. When the engine slows down, they drop inward. Their motion, through linkages, opens or closes a valve that controls the flow of steam (or fuel, depending on the design) into the engine. More speed leads to less input; less speed leads to more input. The machine is made to oppose its own deviation.

If you have ever watched someone steady a bicycle at low speed, you have seen the same principle in flesh. The rider does not decide once, at the start, to keep upright. The rider constantly senses tilt and corrects it, and those corrections change the very state that will be sensed next. The bicycle stays up because the rider is trapped in a rapid loop of measurement and action. The governor was an attempt to trap a steam engine in its own version of that loop.

It is easy, with modern eyes, to treat this as a quaint piece of antique engineering. But the governor was not merely a gadget. It was a proposal about how to get *goal-like behavior* out of matter. Instead of asking for a perfect design that behaves correctly under all conditions, you arrange a system that notices when it is behaving incorrectly and pushes back. The "goal" is not an image in the machine's head; it is the system's tendency to reduce error.

The industrial revolution, famously, was not gentle. Machines were pushed into roles they were never intended to play, then improved in haste, then pushed again. Factories wanted more power, faster production, tighter timing. As engines became stronger and turned faster, governors did something that should have surprised anyone who believed in simple cause-and-effect: they began to misbehave in a way that looked almost psychological. The term engineers used was *hunting*. Instead of settling smoothly into a steady speed, the system would overshoot, then correct too much, then overshoot again, sometimes settling into a rhythmic oscillation: fast, slow, fast, slow. A governor meant to enforce calm would induce a kind of mechanical anxiety.

Hunting is not a minor footnote; it is the "glitch" that taught engineers (and later scientists) what a loop really means. In a linear chain, if something goes wrong you look for the faulty link. In a loop, the parts may each work exactly as designed and the system still becomes unstable. The problem is not a broken piece. It is the *dynamics* of the circular interaction: delays, friction, inertia, and the strength of the corrective action. A governor that reacts too aggressively, or too slowly, can turn the act of correction into the source of error. Purposeful behavior, it turns out, is not merely a matter of having feedback; it is a matter of having feedback with the right timing and gain.

Here is a concrete way to picture it. Suppose an engine is running a little fast. The governor senses the increase and closes the valve. But the engine does not respond instantly. Steam already in the cylinder continues to do work; the heavy flywheel keeps turning; the speed continues to rise for a moment. By the time the speed actually begins to fall, the governor may have already closed the valve too much. Now the engine slows too far. The governor opens the valve, but again the response is delayed, and the correction arrives late and too strong. The governor does not merely damp the disturbance; it amplifies a dance between delay and overcorrection. The loop becomes a pendulum.

This matters for our theme because it marks the moment when "goal-directedness" stopped being mystical and started being technical. You can build a system that seems to want something, and you can also build a system that seems to *panic* when trying to get it. Both behaviors emerge from the same logic: error measurement coupled to action in a loop. The difference lies in the mathematics of stability.

By the middle of the nineteenth century, the governor had become famous enough to attract the attention of James Clerk Maxwell, better known today for unifying electricity and magnetism. Maxwell wrote a paper in 1868 with a title as plain as a workshop manual: "On Governors." What he did in that paper was quietly revolutionary. He treated the governor-and-engine combination not as a collection of parts, but as a single dynamical system with equations describing how it changes over time. He asked not merely, "Does it correct speed?" but "Under what conditions does it correct speed *stably*?"

In other words: when does the machine actually behave as if it has a calm, steady goal, and when does it become a loop that cannot settle down?

Maxwell's move is worth lingering on because it foreshadows a major intellectual pivot. Physics had already learned to write equations for planets and pendulums, but those systems did not *regulate* themselves; they simply followed laws. A governor is different. It is designed to *use information about itself* to change itself. Maxwell's analysis made self-regulation a legitimate subject for mathematical science. It also put "glitches" like hunting into the same category as resonance and instability: not moral failings, not mere annoyances, but lawful consequences of feedback structure.

Once you notice this, the governor becomes a kind of philosophical instrument. It does not just keep a shaft turning at a steady rate. It shows how a system can be organized so that what happens next depends on what just happened, in a way that enforces a pattern over time. The "goal" is no longer something you have to smuggle in as an invisible intention. It is a property of the closed loop.

This is the point where language begins to wobble. People naturally describe the governor as if it *knows* the speed and *decides* to correct it. Even engineers, who understand every pin and bearing, slip into this shorthand because it is so efficient. But it is also dangerous shorthand, because it tempts us to confuse an explanatory convenience for an inner ghost.

The more honest phrasing is that the governor *measures* speed (through centrifugal effects), *compares* it to an implicit desired range (encoded in geometry, weights, and spring tension), and *acts* to reduce the difference. The desired speed is not a number written anywhere; it is the speed at which the forces balance so that the linkage holds the valve in a particular position. Change the weight of the balls or the stiffness of a spring, and you change what the system will settle on. The "setpoint," in modern terms, is built into the physics.

So far this may still sound like engineering, distant from life. Yet this is exactly the sort of story biology forces us to tell, again and again, if we resist the temptation to put a little manager inside every cell. An organism survives because it keeps certain variables within viable bounds: temperature, hydration, blood sugar, salt balance, oxygen. It does so by sensing deviations and acting to reduce them. When those loops malfunction, the organism can oscillate, overcorrect, or collapse. Disease, in many cases, is not a broken part so much as a broken *regulation*. The governor is an early mechanical parable for that deeper truth.

The historical road from the governor to the science of regulation was not straight. It passed through war, telecommunications, and the uncomfortable realization that information can be as causal as force. During the Second World War, engineers and mathematicians grappled with the problem of anti-aircraft fire: how to aim at a moving target when there is delay in sensing, computing, and moving the gun, and when the target changes its behavior in response. This was not a problem that could be solved by a single calculation made at the start. It required continual correction: measure the error, adjust the aim, measure again. The loop reappeared, now in systems that fused sensors, prediction, and action.

From this ferment emerged the field that Norbert Wiener named *cybernetics*, from the Greek word for a steersman. The steersman is a beautiful metaphor because it is not about raw power; it is about correcting course amid disturbances. A ship does not need a stronger engine to stay on course; it needs a control loop linking compass to rudder, perception to action. Wiener and his collaborators argued that the same principles apply in animals and machines: control and communication form a single conceptual fabric. Where earlier thinkers had debated "purpose" as a metaphysical category, cybernetics treated purposeful behavior as a kind of organization that could, in principle, be built, analyzed, and compared across domains.

A key move, made explicit in mid-twentieth-century writings about behavior and teleology, was to separate *teleology* from *teleonomy*. Teleology, in the old philosophical sense, suggests that ends or purposes exist as causes in their own right, pulling events toward them. Teleonomy keeps the everyday language of purpose but empties it of mysticism: a system is teleonomic if it behaves *as if* it has a goal, because it uses feedback to reduce error. The governor is teleonomy in brass and iron.

This does not reduce the wonder; it relocates it. The wonder is no longer that matter can be haunted by intention. The wonder is that a certain geometry of causality—the loop—can create behaviors that look like intention from the outside. The loop is a machine for turning disturbances into information and information into correction.

It is also a machine for making trouble. Once you see hunting as a lawful product of delays and overcorrection, you begin to recognize its cousins everywhere: the shower you cannot set to a comfortable temperature because the hot and cold respond too slowly; the microphone that squeals when placed too near the speaker; the market that booms and crashes as policy lags behind sentiment; the immune system that overshoots into allergy; the body weight that rebounds after dieting. In each case, the "glitch" is not an accident. It is a signature of feedback.

The governor thus delivers a double lesson. First, it demystifies goal-directedness. Second, it warns that goals are not enough. A loop can be purposeful and still fail; in fact, the more strongly it pursues its aim, the more it may oscillate if its corrections arrive out of phase with reality. Regulation has to be tuned to time.

This tuning is not merely a matter of engineering finesse. It is a deep constraint on what any living system can do. A cell cannot respond instantly to a change in its environment; gene expression takes time, proteins must be built, membranes must be remodeled. A tissue cannot repair itself in a heartbeat; cells must migrate, signals must propagate, structures must grow. If the correction arrives too late, it may correct a past problem that no longer exists. If it corrects too strongly, it may create a new problem that then demands correction. The living world is full of loops trying to avoid hunting.

There is one more twist hidden inside the governor, easy to miss if you focus only on its steady operation. The governor does not merely "sense" the engine; it is *coupled* to it. The act of measuring speed is inseparable from the act of being driven by the engine's motion. In modern language, the sensor is not a detached observer. It is part of the system. This is why loops feel uncanny: the system is, in a sense, watching itself by means of itself. That self-coupling is the seed of a much broader idea: that the boundary between "controller" and "controlled" can be porous, and that what counts as a goal can be embedded in the physics of interaction rather than declared from above.

The steam governor was built to solve a practical problem, and it did so well enough to become an icon of industrial stability. But its deeper legacy is conceptual. It helped make it intellectually respectable to treat purpose-like behavior as an emergent property of feedback. It also forced thinkers to confront the fact that loops have personalities: calm, twitchy, sluggish, oscillatory, brittle, robust. These are not metaphors; they are dynamical regimes.

If the first great achievement was to remove the ghost from the machine, the next challenge is subtler: to notice that removing the ghost does not remove the *goal*. The goal persists, not as an idea, but as a set of constraints that the system continuously enforces. The question then becomes: where do those constraints live in biology? In a steam engine, the desired speed is built into ball weights, arm lengths, and valve geometry. In living tissue, the "desired" state might be a body temperature, a salt concentration, a wound closed, a limb the right shape and size. The startling claim of modern biology is that such targets can exist without a central planner.

To see how, we need a new metaphor that is less industrial and more intimate: not the governor spinning on a shaft, but the thermostat quietly clicking in a wall, maintaining a setpoint against the chaos of weather. The thermostat will let us name the hidden ingredient that the governor hints at but does not fully articulate: the *setpoint* itself, the invisible reference value that turns feedback into decision.

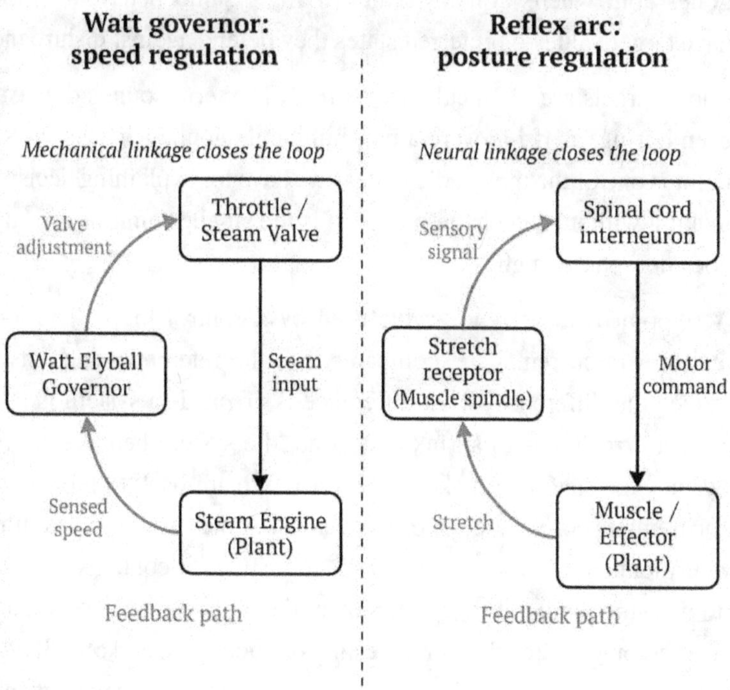

2.2 Invisible Thermostats

The governor taught a blunt lesson: you can get something that looks like purpose out of a loop, and you can also get something that looks like panic. Both are consequences of the same circular logic. The remaining mystery is quieter. A loop can correct, but what is it correcting *toward*? A steam engine has an obvious reference: a desired speed chosen by a designer. The governor's geometry bakes that choice into metal. Biology seems harder. A lizard warms itself on a rock until its body reaches a narrow range, then moves away. A human sweats, shivers, blushes, pants, seeks shade, craves salt, loses appetite when sick, and grows a fever on purpose. A cut on your skin closes, not merely by filling space, but by stopping when the gap is gone. Somewhere, at many scales, living systems behave as if they have target values—preferred states they defend against disturbance.

Those targets are often called *setpoints*. The term sounds domestic, even boring, as if life were a programmable appliance. Yet the setpoint is one of the most radical ideas we have for explaining decision-making without a decision-maker. It is the smallest unit of *goal* that does not require a ghost.

A setpoint is a reference value used by a control loop. The loop measures a current state, compares it to the reference, and acts to reduce the difference. That difference is *error*. If a system is built so that error tends to shrink over time, the system behaves as if it "wants" the state that makes error zero. We are back to teleonomy: apparent purpose made out of mechanics. A thermostat, the familiar wall-mounted kind, is a setpoint made visible. It contains a sensor (to measure temperature), a reference (the number you choose), and an effector (a switch that turns heating or cooling on and off). It does not care about comfort or winter; it simply reduces error between "what is" and "what is set."

The human body is full of thermostats, but most of them do not look like devices. They are distributed across tissues, hormones, ion channels, reflex arcs, and networks of cells that never meet in one place. They are *invisible* in the way gravity is invisible: you infer them from what they do. You can see the body stubbornly returning to certain ranges, as if pulled by an attractor, and you can watch it spend energy to do so. The body does not merely drift toward equilibrium like a puddle seeking flatness. It actively maintains conditions that are not guaranteed by physics alone.

A historical detour helps, because the very idea of "automatic" regulation had to be invented before we could recognize it in ourselves. Long before digital sensors, people built devices that kept temperature stable for practical reasons: incubation, fermentation, metallurgical work, the care of premature infants. One of the earliest stories told in the history of thermostatic control involves Cornelis Drebbel, a Dutch inventor in the early seventeenth century, who demonstrated a furnace or incubator regulated by a temperature-sensitive mechanism. Whether or not every detail of the anecdote survives scrutiny, the cultural fact matters: people began to imagine a machine that could *maintain* a temperature rather than merely *reach* one. Reaching is a one-time event; maintaining is a relationship with the future. It is a promise made by a loop.

By the nineteenth century, thermostats became infrastructure. Heating moved from a fireplace you constantly tended to a building system you expected to behave. Warren S. Johnson, working in the 1880s, developed thermostatic systems for buildings that could regulate temperature across rooms using sensors and control signals. The thermostat became a quiet mediator between weather and architecture, translating disturbance into correction. The public learned to trust an invisible setpoint: you asked for 20 degrees Celsius (or 68 Fahrenheit) and expected the building to cooperate.

Once you have lived with such devices, you carry the metaphor into biology almost without noticing. We speak of a "normal" body temperature as if "normal" were a setting on a dial. We talk about "regulation" of blood sugar, "control" of breathing, "balance" of electrolytes. What the metaphor hides is that these are not single dials. They are nested, sometimes competing loops, operating with different sensors, different time scales, and different costs.

Begin with a case where the setpoint idea fits almost too well: core body temperature. Humans keep their internal temperature in a narrow range because the chemistry of life is fussy. Enzymes speed up or slow down with temperature; membranes become too rigid or too leaky; proteins misfold. In a cool room, your skin may chill, but your core is defended. Blood vessels constrict to reduce heat loss. Muscles generate heat by shivering. Hormones adjust metabolic rate. You reach for a sweater, and that behavioural choice is itself part of the loop. All these actions make sense if the body is continually computing an error: error $= T_{\text{set}} - T_{\text{core}}$. When error is positive (too cold), heat-producing and heat-conserving actions increase; when error is negative (too hot), sweating and heat-dissipating actions increase.

It is tempting to picture this as a single thermostat in the brain. There is certainly a great deal of central coordination. But the setpoint concept does not require one central knob. It requires a reference signal somewhere in the system, and the ability to compare against it. Many comparisons can occur in parallel, and what looks like a single "temperature setpoint" can actually be an emergent compromise between multiple loops.

The simplest thermostat is a two-state switch: on when below setpoint, off when above. That design works, but it oscillates. You feel it in older heating systems: the room cools, heat blasts on, the room warms, heat shuts off, repeat. A smoother controller uses graded responses, adding more heat the farther you are from the target, but even that can overshoot if delays are large. The governor's hunting returns, now as the familiar annoyance of a shower that alternates between scalding and cold because the plumbing responds too slowly to your corrections. The deep lesson is the same: stability is an achievement, not a given. Life achieves it not with one perfectly tuned loop, but with many loops layered and buffered against one another.

The power of the setpoint idea becomes clearer when you look beyond temperature to variables that do not feel like "thermostats" at all. Blood glucose is a classic example. After a meal, glucose rises. The pancreas responds by releasing insulin, which helps cells absorb glucose and tells the liver to store it. Between meals, glucose begins to fall. Other hormones counteract insulin's effects, prompting the liver to release glucose and reducing uptake in some tissues. The body defends a range, not a single number, and it does so because too much glucose damages tissues while too little deprives the brain of fuel. What looks like "metabolism" is also a setpoint-maintaining loop. Diabetes, in this framing, is not merely a chemical problem but a control problem: the loop's ability to sense, respond, or coordinate is impaired, so the system cannot keep error small.

Now consider an example that feels almost uncanny in its goal-directedness: wound healing. When you cut your skin, the body does not merely pour material into the wound until it runs out. It stops. The stopping is the interesting part, because it implies a target state: *closed.* Cells migrate, proliferate, lay down extracellular matrix, contract the wound edges, remodel the scaffold. If the body simply "added tissue," you might expect runaway scarring. In reality, the process is regulated. Sometimes it fails, producing chronic open wounds; sometimes it overshoots, producing hypertrophic scars or keloids. Those failures look, in the language we have been building, like mis-set setpoints or mis-tuned loops: the system either cannot reach the target or cannot recognize that it has reached it.

The setpoint idea becomes even more provocative when you move from physiology to anatomy. A salamander that regrows a limb does not regrow an arbitrary blob of cells. It regrows an arm with the correct shape, proportion, and orientation. How does tissue know when an arm is the right length? Where is the ruler? Developmental biology has spent a century trying to answer versions of that question, and the answers increasingly point to something that resembles a distributed control system. Chemical gradients, mechanical forces, and bioelectric patterns act as signals of "where you are" and "what you should become." Cells make local decisions based on local information, yet the result is a coherent, scaled structure.

This is the point where the setpoint stops being a metaphor and becomes a hypothesis about information in living matter. If a tissue maintains a particular shape, that implies a reference state for that shape, even if no single cell holds the whole blueprint. The reference could be encoded in boundary conditions, in patterns of electrical potential across a tissue, in gene regulatory networks that stabilize certain architectures, or in the mechanical tensions that cells sense through their junctions. The details differ by system, but the logic is shared: there is a preferred state, there is a way to detect deviation, and there are effectors that push back.

A danger lurks here. If every stable outcome is called a setpoint, the concept becomes so broad it explains nothing. A rock at the bottom of a hill is "stable," but no one wants to say the rock has a setpoint. The difference is *active correction*. A setpoint implies that the system expends resources to oppose perturbations. Push a rock and it stays where it lands. Push core temperature and the body shivers or sweats to return. Push a developing tissue and it may reorganize to restore pattern. The hallmark is not stability alone, but stability maintained *against* disturbance by sensing and action.

Once you accept that, a new kind of question becomes natural. Not "what does this organ do?" but "what variable is this loop defending?" The question has a faintly psychological flavour, because it frames organs and cells as agents with preferences. Yet it can be answered mechanistically. The defended variable is the one whose error drives the loop.

Cybernetics, in the mid-twentieth century, gave this idea a general language. It also gave it a philosophical edge. When Norbert Wiener and his colleagues wrote about purposeful behaviour, they were not trying to romanticize machines. They were trying to remove a confusion: the belief that purpose requires a special kind of cause. Feedback offered a way to talk about purpose in the same breath as physics. W. Ross Ashby sharpened the argument by insisting that regulation is not magic; it is a capacity constrained by what the system can sense and do. A regulator must be able to produce enough responses to counteract the variety of disturbances it faces. A thermostat in a mild climate can be simple; one in a chaotic environment must be more sophisticated, or it will fail.

That brings us to the next layer of the story: setpoints are not free. Maintaining a variable costs energy, and sometimes the best way to survive is not to hold a constant setpoint at all, but to move it.

Consider fever. From the outside, fever looks like a broken thermostat: the body runs too hot. But fever is often a coordinated, purposeful state. The body behaves as if it has temporarily raised the temperature setpoint, then works hard to reach it. That is why fever begins with chills and shivering: you feel cold because your current temperature is below the new reference. Only later, when the setpoint drops back toward normal, do you sweat to lose heat. The subjective experience tracks the error, not the absolute temperature. Fever makes sense as a strategy: higher temperature can inhibit pathogens and enhance certain immune processes. But it is also costly. You burn energy, strain the cardiovascular system, risk damage if temperature rises too high. This is regulation as trade-off, not regulation as perfection.

This is where the concept of *allostasis* enters: “stability through change.” In allostasis, the body does not defend a single fixed reference value. It adjusts reference values in anticipation of needs or in response to chronic conditions. Stress hormones rise before waking, preparing the body for activity. Heart rate and blood pressure shift with posture and threat. Appetite, sleep, and pain sensitivity change with illness and environment. Allostasis does not abandon setpoints; it treats them as movable, negotiated targets, coordinated across multiple systems.

The darker companion concept is *allostatic load*: the long-term cost of running these adjustments too often or too intensely. A thermostat that must constantly battle an open window will wear out the furnace. A body that must constantly mobilize stress responses will pay with inflammation, metabolic disruption, impaired immunity, and mental exhaustion. Seen through this lens, many modern ailments resemble a control system under chronic disturbance, forced into costly compensation.

The setpoint, then, is not only a target. It is an economic commitment. To defend a variable is to spend something: ATP at the cellular level, calories at the organism level, attention and behaviour at the psychological level. The “intelligence” of a living system includes choosing what to defend, how tightly to defend it, and when to relax.

This is the pivot that makes setpoints feel like the fundamental unit of biological decision-making. A decision, in its barest form, is not a conscious choice. It is a conditional commitment: *if error is positive, do this; if error is negative, do that.* The decision is embodied in the feedback policy. A bacterium swimming up a chemical gradient implements a policy. A cell maintaining its internal pH implements a policy. A tissue restoring a pattern implements a policy. None of these need a brain. They need a reference, a measurement, and a way to act.

One more anecdote makes the invisibility of setpoints vivid. If you have ever entered a building in winter and felt a sudden wave of warmth, you may have been grateful to a thermostat you never saw. In older systems, the thermostat did not even switch the furnace directly; it signaled a human operator to open or close dampers. Human-in-the-loop control is a reminder that "automatic" regulation is a spectrum. Biology, too, uses mixed control: some loops are local and fast, others are slow and centralized, and many recruit behaviour as the ultimate effector. You cannot understand a setpoint without asking who or what is allowed to adjust it, and at what time scale.

The practical consequence is that setpoints are rarely solitary. A body temperature setpoint interacts with hydration, because sweating loses water. A glucose setpoint interacts with fat storage, because energy can be buffered. A tissue's shape setpoint interacts with growth signals, because growth is not merely addition but patterning. Each defended variable is part of a web, and the web can only be robust if regulation is distributed.

At this point, the thermostat metaphor has done its job and begun to break, which is exactly what a good metaphor should do. It has shown how purpose can live in a loop and how a target can be real without being conscious. But it also risks suggesting that biology is organized around a single master dial for each variable. Life is messier, richer, and more interesting. The targets are nested. Some loops defend fast, local variables; others defend slow, global ones. Some loops tune other loops, changing gains and shifting references. What looks like a single setpoint is often a compromise produced by many controllers, each with partial information and partial authority.

That nested organisation is not an afterthought. It is how intelligence without brains scales up. If a cell had to understand the whole organism to behave usefully, life would collapse under its own informational burden. Instead, competence is distributed: smaller loops solve smaller problems, and their solutions become the raw material for larger loops. The next step is to see this not as a loose metaphor but as an architecture—a set of "Russian dolls" of control, with feedback inside feedback, each layer making the next possible.

2.3 The Russian Dolls of Control

A thermostat in a hallway is comforting partly because it is lonely. One sensor, one setpoint, one actuator, one room. If you want the room warmer, you nudge the number and the rest follows. Biology rarely grants that simplicity. Even the familiar "temperature setpoint" is enforced by sweating, shivering, blood flow, muscle tone, behaviour, and social habits like clothing and shelter. No single part owns the goal. The control is spread out, and it is spread out for a reason: living systems cannot afford a single point of failure, and they cannot wait for a central committee meeting every time a molecule drifts out of range.

Once you start looking for setpoints, you begin to notice a second pattern that is even more revealing. The loops do not sit side by side like appliances. They sit *inside* one another, nested like Russian dolls. A small loop keeps a local variable steady. That local steadiness becomes the reliable groundwork that a larger loop can treat as "given." The larger loop can then pursue a slower, broader goal without micromanaging every detail. Intelligence, in this view, is not something you bolt onto the top of life once brains appear. It is what you get when control loops stack across scales, each one competent in its own domain.

This nesting is so common that we often miss how strange it is. Imagine asking a factory manager to control every rivet in every machine in real time. Impossible. But this is exactly what a mythical "central controller" would have to do if it were responsible for all of biology. The way out is modular competence: each level of organisation takes responsibility for some errors and passes a calmer, simpler world upward.

The simplest way to feel the logic is to start at the bottom, where "decision-making" seems least plausible: molecules.

Inside a cell, concentrations of ions and metabolites are constantly being disturbed. Water leaks, pumps work, reactions consume and produce chemicals, proteins fold and unfold, membranes stretch and relax. Yet the cell maintains a recognisable internal state for hours, days, sometimes years. This is not equilibrium in the physics-textbook sense, because the cell is not a closed container that drifts toward stillness. It is an *open* system powered by energy, and it uses that energy to hold itself in a corridor of viability. When the corridor narrows, the cell dies.

Consider a classic molecular loop: regulation of gene expression. A gene is not merely "on" or "off." It is part of a circuit. The gene is transcribed into RNA; RNA is translated into protein; the protein can feed back to regulate the gene that produced it. Often that feedback is negative: when there is enough protein, the gene is suppressed; when the protein falls, the gene is released. That simple loop stabilises a concentration despite noisy production and degradation. Here the setpoint is not a dial but a balance point created by binding affinities, degradation rates, and the availability of transcription factors. No cell needs to "know" the number of molecules desired. The dynamics embody it.

Now zoom out to the cellular level. A cell must keep not only molecule counts but also *states*: whether it divides, differentiates, migrates, or dies. Those are not single variables; they are patterns of many molecular signals coordinated. A cell has to decide, for instance, whether DNA damage is mild (repair and continue) or severe (halt division or trigger programmed death). The decision emerges from layers of loops: sensors that detect damage, pathways that amplify signals, checkpoints that impose delays, repair systems that reduce the error, and meta-loops that change the sensitivity of the whole system depending on context. A local molecular loop thus becomes a component of a larger cellular loop.

Then the leap that matters for our theme: tissues.

A tissue is a society of cells. Its problems are not simply the sum of cellular problems, because tissues must maintain spatial patterns: boundaries, layers, orientations, proportions. Tissues must make collective decisions such as "close this wound," "regrow this appendage," "stop growing now," or "keep this barrier intact." The astonishing fact is that tissues can do this robustly even though individual cells are short-lived, replaceable, and limited in their perspective. That is only possible if tissue-level goals are not stored in any one cell, but are enacted by distributed loops that coordinate many cells at once.

One way to see nested control in action is to watch a wound heal. At the smallest scale, molecules regulate inflammation, clotting, and cell migration. At the cellular scale, fibroblasts and immune cells respond to signals, change gene expression, and move. At the tissue scale, the wound edges contract and the gap closes. Each scale has its own "error." Molecular error might be "too much reactive oxygen," cellular error might be "wrong phenotype for this environment," tissue-level error might be "gap still open." None of these errors is reducible to the others, and yet the system succeeds because the loops are stacked. If you disrupt only one layer, the system may compensate using others; if you overwhelm the system with disturbance, the loops can conflict and pathology appears.

This brings us to W. Ross Ashby, whose work in the mid-twentieth century gave the nesting idea a rigorous backbone. Ashby was a psychiatrist by training and a cybernetician by conviction, and he had a talent for taking an apparently philosophical claim and turning it into a constraint you could not evade. His most famous formulation is the *Law of Requisite Variety.* "Variety" here means the number of distinct states a system can take, or the number of distinct disturbances it can face. The law, stated informally, says that a regulator must be able to match the variety of the disturbances it needs to control. If the environment can throw a thousand different kinds of trouble at you, and you only have ten kinds of response, you will be overwhelmed.

This is not merely a slogan; it explains why nested control is necessary. A single central regulator would need astronomical variety to counteract disturbances at every microscopic level. But if local loops mop up local disturbances, then the variety that reaches the higher levels is greatly reduced. The world becomes simpler as you climb the hierarchy. The higher-level controller does not need to respond to every molecular fluctuation; it can respond to slower, coarser summaries: oxygen is low, temperature is high, tissue is damaged. Local loops act as "variety sinks," absorbing complexity so that larger goals can be pursued.

Ashby did not leave this as an abstract claim. He built a device to dramatise it: the *Homeostat.* The Homeostat was an electromechanical system designed to maintain stability in the face of disturbance, but with a twist. If the system could not stay stable with its current internal settings, it would alter those settings until it could. In modern language it was a demonstration of *adaptation*: control of control. The Homeostat did not merely correct an error; it changed the way it corrected errors when ordinary correction was failing. It was a mechanical parable for learning, and an early hint that "setpoints" and "gains" can themselves be variables under regulation.

This matters enormously for biology. Many living systems are not content with being stable; they are stable *across* change. A developing embryo must keep its trajectory despite noisy gene expression and variable environments. An immune system must learn what is "self" and "not-self" without attacking the body. A tissue must keep its pattern while cells divide, die, and move. Achieving that kind of robustness requires meta-control. When things go wrong, the system does not merely push harder; it may change strategy.

The idea that control is recursive was later given an organisational face by Stafford Beer's *Viable System Model*. Beer was interested in how organisations survive, but his model is strikingly biological: systems within systems, each viable in its own right, each with channels for communication and control. A liver cell is a viable system in miniature; the liver is a viable system; the organism is a viable system; a colony or society may be one too. Beer's key claim was that viability requires recursion: each subsystem must have some autonomy and regulation, or else the whole becomes brittle. Even if you never apply Beer's model directly to tissue, the intuition is right: life is not a single conductor leading passive instruments. It is a nested ensemble where each instrument can tune itself.

At this point, a reasonable objection arises. If control is everywhere, is "intelligence" just a synonym for "feedback"? If so, the concept risks becoming too generous. A pond has feedback between evaporation and humidity; a thermostat has feedback; a cell has feedback; a brain has feedback. Have we explained anything?

The way out is to pay attention to *competency*: what problems can the system solve, across what range of disturbances, and how flexibly can it change its behaviour when the usual tactics fail? A thermostat can regulate one variable in a narrow range. A homeostat can adjust its own parameters to regain stability across a wider range. A tissue can sometimes rebuild an anatomical structure after injury, implying a richer internal organisation of goals and error signals. Nested control is not merely a repetition of the same idea; it is the means by which competency scales up without centralisation.

A second objection is subtler and cuts closer to our theme of "ghosts." When a system behaves as if it has a goal, it is tempting to say it must have a representation of that goal, a model of the world, a blueprint. In engineering, you can often point to such a model explicitly: a thermostat "represents" the setpoint as a number you dial in. In biology, where is the representation? If no cell contains the full blueprint for a limb, how can a limb be rebuilt?

Here the tradition sometimes called the "good regulator" idea becomes useful. The slogan version is that any effective regulator must embody a model of the system it regulates. That can sound like a return of the ghost: if regulation requires a model, and models live in minds, then perhaps we smuggled intelligence back in. But a "model" in this context does not have to be a picture or a sentence. It can be a structure whose behaviour mirrors relevant aspects of the world. A spring "models" force and displacement; a flyball governor "models" speed; a set of coupled bioelectric gradients can "model" anatomical polarity by providing a stable reference frame. The model is not necessarily symbolic. It is a constraint made physical.

Nested control makes this less mysterious. A higher-level loop does not need a detailed model of molecules; it needs a model of the aggregate behaviour that lower loops deliver. In other words, it can treat a cell as a reliable module: a thing that maintains its own internal setpoints, responds within a predictable range, and provides stable interfaces to its neighbours. The "model" is distributed across levels: each level embodies what it needs to embody to regulate its own errors, and the composition yields competent behaviour at larger scales.

A vivid biological example of nested control is the regulation of blood pressure. At the molecular level, receptors in blood vessels detect chemical signals and mechanical stretch. At the cellular level, smooth muscle cells constrict or relax. At the organ level, the kidneys regulate blood volume and salt balance. At the whole-organism level, the nervous system adjusts heart rate and vessel tone depending on posture, exertion, threat, and sleep. If you stand up quickly, gravity pulls blood downward, pressure at the brain drops, and you may feel dizzy. Within seconds, reflex loops tighten vessels and increase heart rate. Over hours and days, kidney regulation adjusts volume. If any layer is impaired, others may compensate for a while, but the cost can be high. Chronic compensation is a kind of allostatic load: the system is still regulating, but it is doing so in a strained, expensive way.

Notice how this reframes agency. The organism does not "decide" blood pressure by deliberation. It enacts a policy distributed across loops. Some loops are fast and local; others are slow and global. The resulting behaviour can look purposeful, adaptive, even prudent, without any one locus of prudence. This is what "intelligence without brains" begins to mean in concrete terms: control over outcomes achieved by architectures of nested feedback.

The nesting also explains why evolution can build complex competencies without requiring a miracle of coordination. Natural selection does not need to invent a perfect organism in one step. It can improve local controllers first. A mutation that makes a membrane pump more reliable reduces intracellular disturbances; that stabilises the cell; stable cells make reliable tissues; reliable tissues make viable organisms. Each improvement is a new doll that can fit inside a larger one.

The same logic explains why organisms can be both robust and fragile. Robust because redundancy and layering allow compensation; fragile because loops can interfere. A strong tissue-level growth signal may conflict with cellular-level damage checkpoints, producing cancer when the balance fails. A powerful immune loop may overshoot and attack the body. A developmental loop that normally corrects pattern may, under unusual disturbances, lock into a stable but wrong attractor. In other words, intelligence has failure modes, and nested control gives you not just stability but a taxonomy of ways to go wrong.

Second-order cybernetics pushes the nesting one step further by insisting that the observer is not outside the loop. Any time we say a system has a goal, we are describing it relative to a set of variables we chose to measure. This is not a dismissal; it is a discipline. It reminds us that "control" is always control of some variable, and the choice of variable matters. In the laboratory, you might treat a tissue as regulating voltage patterns; in the clinic, as regulating wound closure; in evolution, as regulating reproductive success. These descriptions are not mutually exclusive. They are different levels of the same nested architecture, viewed through different windows.

Seen this way, the Russian dolls are not only biological; they are also conceptual. We, too, are nested observers: we describe molecular regulation with one vocabulary, cellular behaviour with another, tissue morphogenesis with another, and psychology with yet another. The temptation is to assume these are separate kinds of causation. Cybernetics suggests a more continuous picture: similar loop-logics repeated at different scales, each acquiring new degrees of freedom and new costs.

If all of this still feels abstract, return to the simplest human experience of nested control: learning to balance on a bicycle. Early on, the task feels impossible because you are trying to consciously control everything. With practice, the control sinks downward. Fast reflexes and muscle synergies take over, freeing your attention for traffic and navigation. The higher level stops micromanaging because lower levels become competent. The result feels like intelligence, and it is, but it is also the delegation of control to nested loops. Biology does the same thing without asking permission: it offloads problems downward whenever it can, because downward loops are faster and cheaper.

The most important consequence for the rest of this book is that once you accept nested control, you stop looking for the single "seat" of biological intelligence. You begin instead to ask how goals are partitioned across scales, how conflicts are negotiated, and how information flows between loops so that local competency serves global viability. A tissue that can rebuild a pattern is not a puppet of genes; it is a collective of cells with a shared regulatory landscape, capable of making corrections that no single cell could specify.

At the end of the day, the Russian dolls metaphor is not about hierarchy for its own sake. It is about *compression*. Each layer compresses the messy variety below into a manageable set of signals and actions. That compression is what makes higher-level agency possible without higher-level omniscience.

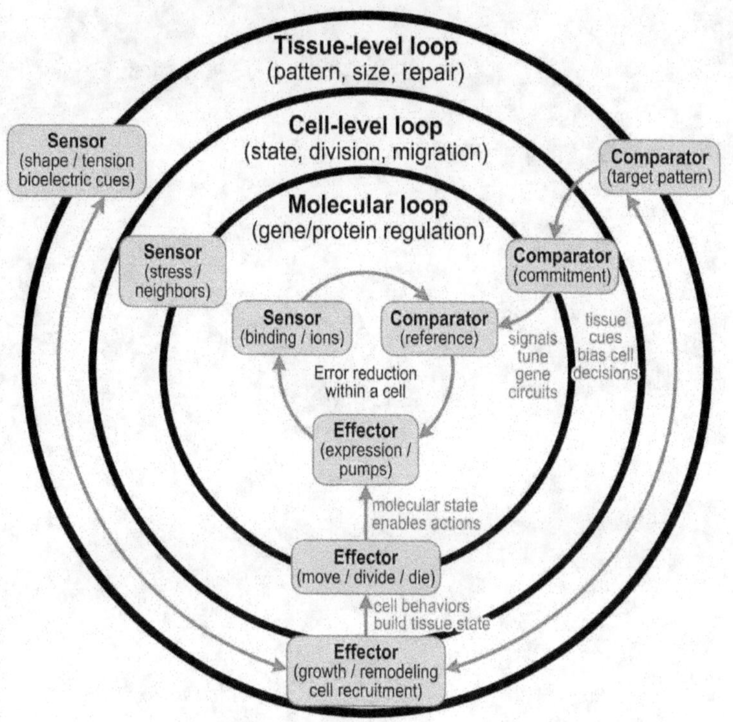

Nested loops compress complexity: local regulation makes higher-level goals feasible.

Nested control leaves a final, unsettling thought. If intelligence is a stack of loops, then the most important events in life may not be the dramatic decisions we notice, but the quiet negotiations among layers we never feel: molecular circuits stabilising cells, cells stabilising tissues, tissues stabilising bodies. The next step is to ask what happens when these layers disagree—when the "goals" of one scale do not neatly serve the goals of another. That is where the story of intelligence without brains becomes not only a story of problem-solving, but a story of conflict, compromise, and the strange politics of living matter.

Chapter 3

The Cognitive Speck

Long before the first neuron fired, life was already solving complex problems of survival in a hostile world. By examining the precise navigation of bacteria and the maze-solving prowess of slime molds, we discover that the fundamental machinery of memory, anticipation, and decision-making is woven into the physical fabric of the cell itself.

3.1 Navigating Without a Map

If you have ever tried to find a bakery in an unfamiliar city using only smell, you have met the central problem of microscopic navigation. You can tell when the air carries more cinnamon and warm yeast, but you cannot easily point to *where* the bakery is. Wind swirls. Odours pool in courtyards and vanish at street corners. Now shrink yourself down to the size of a bacterium and replace the city with a watery world where molecules jitter and smear by diffusion. The challenge becomes harsher still: the creature is not merely without a map; it is without the very notion of a map. There is no bird's-eye view in a cell.

And yet bacteria find food. They do it reliably enough that their success can be measured, graphed, and predicted. In a drop of pond water, in a human gut, in the soil around plant roots, single cells make their living by moving up invisible hills of chemistry and away from chemical cliffs. This is the first place where "intelligence without brains" stops being a slogan and becomes an engineering fact: a bacterium, too small to compare left versus right in any straightforward way, still extracts *direction* from a world that only delivers *moments*.

The crucial trick is that a bacterium does not navigate like a hiker. A hiker looks around, picks a heading, and walks with intention. A bacterium such as *E. coli* swims, stops, and reorients. It alternates between two modes. In a *run*, it drives itself forward in something like a straight line. In a *tumble*, it loses that straightness, briefly thrashing in place and emerging pointed somewhere else. If you watched one under a microscope without any context, you might think it indecisive, even panicky: go, jitter, go, jitter. But seen over time, the pattern begins to read like a strategy.

The historical anecdote that matters here is not a tale of a lone genius having an idea in a single afternoon, but of patience made mechanical. In the early 1970s, Howard Berg and Douglas Brown used painstaking three-dimensional tracking to quantify how *E. coli* moves. Instead of relying on impression, they turned the bacterium into a set of time-stamped coordinates. With that, "run" and "tumble" became measurable events rather than metaphors. What emerged was a remarkably clean picture: the bacterium does not swim faster when it likes what it senses; it changes how often it tumbles. Its speed stays roughly similar. What shifts is the *statistics* of its choices: longer runs in favourable conditions, more frequent reorientations in unfavourable ones.

That difference sounds minor until you imagine living inside it. Steering, in the way we mean it, would require the cell to know its orientation relative to the gradient, to compute "left is better than right," and to adjust its propulsive machinery continuously. But the bacterium does not need any of that. It needs only to decide whether to keep going *as it is* or to roll the dice again. A run is a commitment to the current direction for a little while. A tumble is a request for a new sample of the world, drawn from a distribution of possible headings. In other words, the bacterium turns navigation into a controlled random walk: not a map-following plan, but a probability machine that can be biased.

Randomness here is not the enemy; it is the raw material. At the microscopic scale, water is not a smooth medium but a noisy crowd. A bacterium swims at a low Reynolds number, where inertia is irrelevant and friction dominates; it cannot coast. If it stops beating its flagella, it stops moving, almost immediately. At the same time, the chemical cues it cares about are not stable signposts. Molecules arrive at receptors in a jittery sequence, like raindrops on a roof. You do not measure a drizzle by counting individual drops for one second and declaring it "light" or "heavy." You average. You smooth. You compare now to a moment ago. That is already the shape of a computation.

Here is where the phrase "single cells perform calculus" earns its keep, but only if we handle it carefully. No bacterium writes equations. No bacterium contemplates limits. What it does instead is a physical approximation to a derivative: it compares concentration over time. Spatial derivatives (how concentration changes from left to right) are hard for a tiny cell because its body is so small that there may be almost no difference across it. Temporal derivatives (how concentration changes from one moment to the next) are available because the cell is moving. Motion converts space into time. If the bacterium swims into a region with a higher concentration of attractant, the number of molecules it detects tends to increase over the next few seconds. If it swims the wrong way, detections tend to decrease. In both cases, the cell is sampling a gradient, but the gradient is experienced as a trend.

That trend is the only "compass needle" the bacterium needs. When things are getting better, it delays tumbling; it extends the run. When things are getting worse, it tumbles sooner; it shortens the run and tries a new heading. Over many cycles, those tiny adjustments produce a drift up the gradient. Not because the bacterium knows where it is going, but because it is slightly more willing to keep going when its recent past suggests improvement.

A useful way to picture the logic is to imagine walking blindfolded up a hill with a barometer that reads air pressure. You cannot tell which direction is uphill by looking, but if you take a few steps, you can tell whether pressure is dropping. If it is dropping, you are likely going up; keep going a little longer. If it is rising, you are likely going down; change direction sooner. Repeat. You will not march straight to the summit, but you will spend more time in steps that help than in steps that hurt. Over time, that bias is enough.

The elegance of this strategy is that it solves two problems at once: the problem of direction and the problem of noise. Because the cell bases its decision on a short history rather than on a single instantaneous reading, it is less vulnerable to random spikes and dips. And because it only needs a *sign* (better or worse), it can operate without an absolute reference scale. A fancy map would be fragile; the bacterium's method is robust.

It also exposes something broader about intelligence in living systems: what matters is not representation in the head, but control in the world. The bacterium controls one knob—tumble frequency—and that knob, tuned by feedback, is enough to reshape a random walk into a purposeful search. If we want a more general definition of agency that does not smuggle in consciousness, we could do worse than this: an agent is something that can change the distribution of its future states in a goal-directed way. *E. coli* does not "intend" in the human sense, but it does *act* in the sense that it shifts its probability of being near food a few minutes from now.

The strategy is especially striking when you remember what is missing. There is no central processing unit, no executive decision-maker. The "memory" required to compare now with a few seconds ago is biochemical: transient changes in receptor sensitivity and downstream signalling pathways. The cell builds a short-lived internal state, just long enough to say, "relative to my recent past, things improved," and then resets. This is a tiny memory, but it is a real one. Without it, there is no temporal derivative; without a derivative, there is no way to tell improvement from decline.

At this point it is tempting to treat the bacterium as a miniature mathematician, but that would mislead us. The bacterium is not computing an abstract slope in a vacuum; it is implementing a feedback loop in messy water. The strength of the story is not that biology secretly imitates our textbooks. It is that the structure of the problem—finding a direction in noise without a map—forces any successful solution to resemble certain mathematical ideas. When you must extract a trend from fluctuations, you will, in some form, differentiate and average. Calculus is not a human invention laid on top of nature; it is one of the languages that describes what nature already does when it solves problems under uncertainty.

The random-walk perspective makes the consequence visible. If a bacterium tumbled at fixed intervals no matter what, it would perform a true random walk. It would spread out from its starting point like ink in water, with no preference for food or safety. Add the tiniest bias—just a small increase in the average run length when the recent chemical trend is favourable—and suddenly the cloud of possible positions drifts. The bacterium does not need perfection. It needs only to be a little less wrong than chance, consistently.

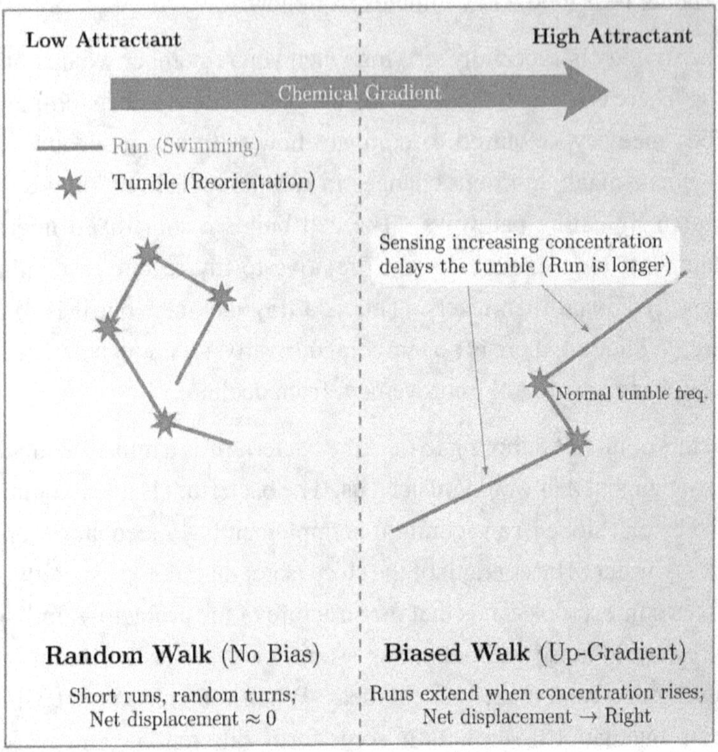

The diagram is a simplification, of course. Real trajectories are three-dimensional and buffeted by currents; real gradients have wrinkles. But it captures the core idea: *the bacterium does not aim; it biases.* In the left panel, each tumble is a fair lottery: any direction is as good as any other, because the cell does not change its rules. In the right panel, the lottery is rigged by experience: recent improvement buys a longer run; recent decline triggers a quicker reorientation. If you were trying to explain this at a dinner party, you might say: the bacterium is not clever about choosing; it is clever about *staying*.

This also clarifies a subtle point about what the cell can and cannot know. Suppose the bacterium happens to be facing up-gradient, and the concentration is rising. It will run longer and thereby move further up-gradient, which looks like success. But suppose it is facing cross-gradient. The concentration may not change much, so the run length may remain average; the cell will not be "punished," but neither will it be strongly rewarded. Now suppose it is facing down-gradient. Concentration will tend to fall; tumbling will come sooner; down-gradient runs will be shorter. Over time, down-gradient headings are sampled less, not because the cell avoids them in advance, but because they fail to earn time.

The method has a built-in humility. The bacterium never assumes it is right. It never commits forever. Every run is provisional, and every tumble is a chance to update. That is an oddly modern philosophy for a creature with no nervous system: behave in a way that can exploit good news without being trapped by bad luck.

Once you see it, the strategy echoes in places far from bacteria. Search algorithms in computer science, from simple hill-climbers to more sophisticated methods that blend exploration and exploitation, are variations on the same theme: keep what improves your score; randomize when you are stuck. Even human behaviour in uncertain environments often follows this rhythm. We try something; if it seems to work, we keep doing it; if not, we change. The difference is that we narrate our choices. The bacterium does not.

All of this raises a question that will matter more as we move to larger, squishier organisms. What counts as "memory" in a being with no brain? In *E. coli*, the memory is brief and mostly internal, a fading trace that lasts seconds. It is enough to compute a temporal derivative, but not enough to build a history. Yet even this minimal memory hints at a general principle: to behave intelligently in time, you need some way to let the past lean on the present. The past can be stored in molecules, in electrical states, in mechanical tension, or even outside the body in the environment itself.

That last possibility sounds odd until you meet a different kind of creature: the slime mould, a living sheet of jelly that can leave reminders of where it has been not in synapses but in the world. The bacterium's navigation shows how far you can go with almost no memory at all; the slime mould will show what happens when memory becomes a physical trail.

3.2 The Memory of Jelly

A bacterium can afford to be forgetful. Its world is fast, its needs are immediate, and its solutions are short-lived: a few seconds of biochemical "recent history" is enough to decide whether to keep swimming or to roll the dice again. The next creature in our story looks, at first glance, even less equipped for cognition. It is not a swarm of cells with division of labour, and it is not a neuron-rich animal with dedicated sensory organs. It is, in a certain unflattering sense, a moving stain.

Physarum polycephalum is often introduced as "the slime mould that solves mazes". The phrase has become a kind of party trick for biology, the sort of fact that gets passed along with a sense of delighted disbelief, ever since the shortest-path experiment reported by Nakagaki and colleagues in 2000. But the risk of a party trick is that it turns a serious phenomenon into a novelty. The deeper point is not that *Physarum* can win a puzzle designed for humans. The point is that it forces us to rethink what memory can be made of. In *Physarum*, memory is not primarily a private inscription inside a brain. It is a public footprint, a change in a material network, a pattern of thickened tubes and abandoned corridors, sometimes even a residue left behind in the environment like chalk on a pavement. It is memory you can see.

To meet *Physarum* properly, you have to set aside the everyday idea of an organism as a compact body with clear edges. In its foraging form, *Physarum polycephalum* is a single cell with many nuclei, spread out as a branching web of living tubes. Inside those tubes, cytoplasm streams back and forth in rhythmic pulses, carrying nutrients and chemical signals. The organism is constantly reshaping itself: extending thin exploratory tendrils, retracting them, thickening a route that pays off, abandoning one that does not. A brain, for humans, is a specialised place where the world is represented. For *Physarum*, representation is distributed across its entire body plan. The "thought," if we dare to use that word, is written into the plumbing.

One of the most famous demonstrations begins with a simple setup: a maze, a damp surface, and food placed at two points. The organism is placed near one opening. At first, it does what any prudent explorer does in a dark building: it spreads out, probing multiple corridors. You could call this stage wasteful, but it is better understood as buying information. Then a second stage follows, and it is the one that makes people sit up: the organism retracts from most of the maze and leaves behind a single thick connection between the food sources, a path that is close to the shortest route. The maze has been "solved" not by a sudden insight but by a physical editing process. Exploration creates many possibilities; success stabilises a subset; the rest are pruned.

There is a historical irony here. Maze-solving is often treated as a triumph of animal intelligence, a classic assay in psychology laboratories. Rats in mazes were once our cultural shorthand for learning. When a slime mould does something analogous, it feels like a category error, which is precisely why the result is so useful. It breaks the association between "maze" and "brain". A maze is simply a network with constraints. If the organism itself is a network that can adapt its connections, then the problem and the solver begin to look strangely compatible.

The adaptation has a concrete physical logic. Tube segments that carry more flow tend, over time, to become thicker and more conductive; segments that carry less flow tend to thin and disappear. The cell is not consciously "choosing" a shortest path. Instead, it is running a feedback loop: flow reinforces conductance, conductance permits more flow. Meanwhile, detours that add length and resistance lose out. What emerges is a kind of embodied optimisation, as if the organism were performing a distributed vote with its own bloodstream. Each branch competes not through argument but through hydraulics.

This is already a kind of memory. The thickened tube is a record of past usefulness. It is a physical persistence that makes the future different from the past. After the pruning, the organism no longer needs to keep "thinking" about the maze; the maze has been inscribed into its morphology. In brains, memory is often imagined as a stable pattern of connections among neurons. In *Physarum*, memory really *is* a stable pattern of connections, but without the neurons. If you wanted a slogan, it would be: the cell remembers by becoming.

The most provocative popular stories compare *Physarum*'s networks with human engineering, especially transport systems. There is a reason this comparison lands emotionally. A subway map represents collective intention: budgets, politics, commuters, and a desire to move bodies through a city efficiently. To see a slime mould produce a network with similar features is to feel, briefly, that efficiency is a natural form, something the world wants. But the sober lesson is more interesting than the romance. Both systems are solving similar constraint problems: connect multiple valuable nodes while keeping costs low and maintaining robustness to damage. In a city, the "cost" is concrete and steel; in *Physarum*, the cost is living tissue that must be maintained, fed, and defended. The resemblance is not mystical; it is convergent. Given similar trade-offs, different substrates can settle on similar designs.

At this point, it helps to borrow a distinction from everyday life. There is the kind of memory you carry in your head, and there is the kind you store in the world: a calendar on the wall, a grocery list on the counter, footprints in snow that tell you which way you walked. The second kind feels almost like cheating, but it is in fact one of the most powerful human strategies. *Physarum* does something analogous, and it does it in a way that is startlingly literal.

As it moves, *Physarum* lays down a thin layer of slime. For a long time this would have seemed like mere mess, the biological equivalent of crumbs. Yet experiments show that this residue matters. When faced with a choice of where to go next, the organism tends to avoid areas it has already explored, as if the slime were a "do not revisit" sign. It is not that the slime has to contain a detailed record. It can be simple: a chemical difference between fresh territory and visited territory. The remarkable part is what that simplicity accomplishes. By marking explored regions, the organism reduces redundant search. It gains the practical benefit of memory without building a library inside itself.

There is an elegant demonstration of this principle that feels almost cruel in its clarity. If you pre-coat the surface with slime, you erase the contrast between "my trail" and "new ground". In that condition, *Physarum* becomes noticeably worse at tasks that require efficient exploration, such as navigating around obstacles where revisiting the same region wastes time, as Reid and colleagues showed in 2012. The point is not that slime moulds have a mind that can be confused, but that their success depends on a particular coupling between organism and environment. Change the environment's ability to hold a record, and you change the organism's behaviour. The "memory" was never solely inside the creature; it was shared with the world.

This gives us a more general way to talk about cognition without lapsing into mysticism. Sometimes the substrate of memory is not an internal archive but a reversible modification of the surroundings. A trail, a scent mark, a scratched tree, a path worn into grass: biology is full of these external notations. What is unusual about *Physarum* is that the external notation is made by a single cell, and that it is used immediately to guide future action. The cell is both author and reader of its own footprint.

If the story ended here, we might conclude that "memory without neurons" means "leave marks and follow them". But *Physarum* complicates matters further, because it can also do something that seems, on the face of it, impossible for a jelly-like network: it can anticipate time.

The most famous anticipation experiment has the structure of a prank that becomes a philosophical problem. The organism is placed on a surface where it can move steadily. Then it is subjected to an unpleasant condition—a pulse of dryness, cold, or some other stress—at regular intervals. Each time the stress arrives, the organism slows down. That much is unremarkable; even a simple chemical reaction can be slowed by a change in conditions. The twist comes after repetition. Once the organism has experienced several pulses at a fixed rhythm, the pulses stop. And yet, at the time when the next pulse would have arrived, the organism slows down anyway, as if bracing for a blow that never comes, as Saigusa and colleagues reported in 2008. It is not prophecy; it is pattern.

To call this "anticipation" risks anthropomorphism, but avoiding the word entirely risks missing what is genuinely strange. The organism's behaviour is not merely reactive to the present. It is shaped by an inferred regularity. It carries forward information about timing. That is, in a plain sense, a memory of intervals.

How could this be done in tissue with no neurons and no clock in the ordinary sense? The most responsible answer is also the least satisfying to our narrative instincts: we do not need a tiny timekeeper; we need dynamics that can entrain. Many physical systems, from pendulums to chemical oscillators, can lock onto a periodic input. If you tap a swing at a regular rhythm, the swing begins to mirror that rhythm, and it can keep swinging for a while after you stop. Nothing about this requires a brain. In *Physarum*, the body already has rhythmic streaming and oscillatory activity. A periodic stressor can couple to those rhythms, shifting their phase and amplitude. After repeated pulses, the internal dynamics may continue for a cycle or two, producing the behavioural slowdown at the "expected" time.

This explanation demystifies without trivialising. The swing is not "thinking", but it does embody a temporal memory: its motion depends on its past. The difference is that in *Physarum*, that temporal persistence is hooked up to a choice that matters to the organism—how fast to move, where to allocate effort, whether to pause. The cell is exploiting a general property of dynamical systems (persistence and entrainment) for a biological purpose (reducing damage). Again, calculus is not required, but something structurally similar is happening: the organism is extracting a parameter of the world (period) and using it to shape action.

At the dinner party, the most honest way to put it might be: *Physarum* does not know the future, but it behaves as though the near future has a rhythm, because the past has impressed a rhythm into its body.

Put the maze-solving, the trail avoidance, and the timing effect side by side, and a pattern emerges. *Physarum*'s "intelligence" is not a single trick; it is the repeated use of persistence. The organism changes in ways that do not instantly disappear. Tubes thicken and remain thick for a while. Slime is deposited and remains on the surface. Internal oscillations can be entrained and then slowly decay.

These are all memories in the most practical sense: they are traces that make the next moment different.

There is also a moral, if we want one, about how easily we confuse the form of a solution with the essence of a solution. When we think of memory, we picture recollection. When we think of planning, we picture a mind rehearsing options. But for living systems, especially small ones, the cheapest "computation" is often a physical change. Why spend energy representing a maze when your body can become a better maze-solver simply by reconfiguring its transport network? Why store a map internally when the ground itself can store your exploration history? Why build a clock when your chemistry already oscillates and can be nudged into a rhythm?

This is not merely a curiosity. It is a way of seeing that scales upward. If a single cell can use its body as a notebook, then tissues can do something similar on a larger canvas: storing information in mechanical stiffness, in electrical potentials across cell membranes, in the distribution of signalling molecules, in the structure of extracellular matrix. Biology repeatedly turns "memory" into a property of material organisation.

There is, however, a limit to what a lone organism can do alone. A slime mould's external memory works because the environment is close and the timescales are manageable. Its network optimisation works because the whole body is the network. But many of the problems life faces are not solvable by a single agent, no matter how clever. Defending an animal body against infection, for example, requires a kind of coordination that no single cell can oversee. The solution is not a bigger brain inside each immune cell. The solution is a society: swarms of cells that communicate locally, amplify signals, and converge on a target together. What *Physarum* shows us is that memory can be written into jelly and slime; what the immune system will show us next is that intelligence can be written into crowds.

3.3 The Immune Swarm

A bacterium biases its randomness. A slime mould writes its past into slime and tubing. Both are reminders that "thinking" can be built from feedback and persistence rather than from a control room in the skull. The immune system adds a further twist. Here the problem is not finding dinner or remembering a corridor; it is survival inside a body that is itself a buffet.

Every minute, microbes land on your skin, enter with the air you breathe, ride in on food, and exploit tiny abrasions you never notice. They are not all enemies, but the immune system must treat uncertainty as dangerous. It must detect trouble quickly, decide what kind of trouble it is, and concentrate force at the right place without destroying the host in the process. If this sounds like the work of a commander, it is. What is striking is that no such commander exists.

The immune response that saves your life on an ordinary Tuesday is, in many cases, a triumph of collective behaviour. Single immune cells are competent—they can sense chemicals, crawl, engulf, kill, secrete signals. Yet the power of immunity lies in what happens when thousands of these cells coordinate their local decisions. Their "intelligence" is swarm intelligence: organisation that emerges from many agents obeying rules that are simple enough to fit inside a cell.

To see this in action, imagine a splinter. The injury is small, but it is also a breach in a carefully maintained border. You might feel the sting and then forget about it. Your immune system does the opposite: it notices the breach and becomes, briefly, obsessed. Within minutes, cells begin to arrive, not politely in a queue but as a crowd that seems to know where to go.

A useful starting character is the neutrophil, a type of white blood cell that acts like emergency services. Neutrophils are fast. They can squeeze through tissues, change shape, and respond to chemical cues. Under a microscope, they do not glide; they crawl with purpose, constantly extending and retracting protrusions, as if tasting the ground with their whole body. Each neutrophil is a "cognitive speck" in its own right, but the more astonishing behaviour appears when they act together.

In living tissue, neutrophils can form swarms at sites of cell death or infection, as demonstrated by Lammermann and colleagues in 2013. "Swarm" is not poetic licence. It describes a coordinated accumulation with distinct phases: a few cells arrive first, then many more are recruited, and finally a dense cluster stabilises around the target. The scene resembles insects converging on a dropped piece of fruit, except the "fruit" is a danger signal and the insects are cells inside your own flesh.

One of the challenges in appreciating this is that, for much of scientific history, immunity was studied in ways that flattened space and time. Cells were mixed in dishes. Chemicals were added. Outcomes were measured in bulk. These approaches revealed a great deal, but they made it hard to see the immune system as a spatial problem: a navigation problem in a complex terrain. The turning point came with methods that allowed researchers to watch immune cells in real tissues of living organisms, a kind of documentary filmmaking at the cellular scale. What these films show is not a tidy march but a self-organising crowd.

The key story, experimentally, is about signals that recruit signals. In a swarm, the first responders are not merely individuals doing their own jobs; they become broadcasters. They release chemicals that call in reinforcements and help align the movements of others. Among these chemicals, leukotriene B4 (often abbreviated LTB4) has emerged as a crucial amplifier in neutrophil swarming. Think of it as a relay flare. A few neutrophils detect danger and release

LTB4, which spreads locally and attracts more neutrophils, which release more, creating a positive feedback loop that concentrates the response. At the same time, adhesion molecules such as integrins help cells stick and stabilise the growing cluster, turning a transient gathering into a persistent structure.

This is a familiar motif now. The bacterium did not steer; it biased. The slime mould did not recall; it left traces and remodelled its network. The immune system does not appoint a leader; it creates gradients, relays, and stickiness that make leadership unnecessary. A swarm is what you get when local rules are tuned so that the group, as a whole, behaves as though it has a goal.

The "goal" in this case is not food but containment. A wound is not only an opening; it is a source of information. Damaged cells release molecules that healthy, intact tissue keeps hidden. Microbes bring their own molecular signatures. The immune system interprets these as cues. But cues have to travel, and travel takes physical form: diffusion, flow, binding, decay. The immune system, like the bacterium, lives in a world of gradients.

There is a particularly vivid example of a gradient that acts like an alarm bell. When tissue is wounded, levels of hydrogen peroxide can rise rapidly at the wound margin and form a spatial gradient that extends across a surprisingly large distance on the scale of cells, as shown in zebrafish by Niethammer and colleagues in 2009. Hydrogen peroxide is a household chemical associated with disinfectant fizz, which makes it an unexpectedly domestic protagonist in the story of immune navigation. In tissues, it becomes part of a signalling field: a diffuse "smoke" that points immune cells towards damage. Chemically, this is often written as H_2O_2.

What matters here is not the specific molecule, but the architecture: the wound generates a field, and cells read the field. The field is not a message addressed to particular recipients; it is a property of space. Any cell equipped with the right sensors can respond. This is one of nature's most economical designs. Rather than telling each immune cell, individually, where to go, the tissue itself is changed in a way that makes the right direction more likely to be chosen.

If you want an everyday analogy, think of a crowd in a building during a fire. A public-address system can give instructions, but smoke and heat gradients also guide behaviour: people move away from heat, towards clearer air, and towards the light of exits. No single person has to be informed in detail. The environment becomes informative. In the immune system, danger makes the tissue legible.

The fascinating part is that immune cells are not guided by a single attractant. They integrate many cues, and those cues can conflict. Some signals are produced by microbes, some by damaged host cells, some by other immune cells. Different receptors respond to different molecules, and the cell must decide how to weight them. This is not deliberation in the human sense, but it is decision-making in the control-theory sense: combining inputs to produce an action. When the immune response is appropriate, we call it protection. When it is miscalibrated, we call it inflammation or autoimmunity. The difference often comes down to how these local decision rules are tuned.

There is also an emotional dimension that makes the immune swarm more than an abstract machine. The immune system protects us by being, in some ways, willing to damage us. Swarming neutrophils release enzymes and reactive molecules that kill microbes but can also harm tissue. The redness, heat, and swelling of inflammation are not incidental; they are consequences of a defensive strategy that trades local discomfort for systemic survival. Swarm intelligence is not gentle intelligence. It is effective intelligence under pressure.

At this point, it is tempting to draw a neat boundary: bacteria do short-term navigation; slime moulds do physical memory; immune systems do collective defence. In reality, the categories bleed into one another. Neutrophils, like bacteria, navigate by gradients. The immune system, like slime moulds, uses the environment as a substrate for information. And, perhaps most surprisingly, parts of immunity exhibit something that looks like memory even when the classical "memory system" of adaptive immunity is not involved.

Most people learn, at some point, a tidy dichotomy: innate immunity is fast and nonspecific but forgetful; adaptive immunity is slower but specific and remembers, which is why vaccines work. The tidy story is useful, but nature does not always respect tidy stories. In recent years, evidence has accumulated that innate immune cells can undergo long-lasting functional changes after certain infections or vaccinations, changes that alter how they respond to later challenges even when those later challenges are unrelated. This phenomenon is often called *trained immunity*.

The word "trained" is provocative for the same reason "anticipation" was provocative in the slime mould. It invites the image of learning. Again, the sober version is still astonishing: the cell's future behaviour changes because its gene expression programs have been reconfigured by experience, as described by Netea and colleagues in 2016. Not by changes in DNA sequence, but by changes in the regulatory machinery that determines which genes are accessible and how strongly they are expressed—epigenetic modifications and shifts in cellular metabolism that can persist. The cell, in effect, carries a memory not as a detailed record of a particular pathogen, but as a durable change in readiness.

This is not adaptive immunity's exquisite specificity, with antibodies tailored to particular targets. It is something rougher: a heightened or altered state that can make a later response stronger or faster. One of the contexts in which trained immunity has been discussed is the observation that some vaccines, such as BCG (originally developed against tuberculosis), appear in some studies to have broader effects on susceptibility to other infections. The details and the degree of these effects are debated and context-dependent, but the conceptual shift is clear: the innate immune system may not be the blank slate it was once portrayed as.

Here, the theme of "memory without brains" returns in a new register. In *Physarum*, memory could be a slime trail or a thickened tube. In innate immunity, memory can be a chromatin landscape, a biochemical posture that makes certain responses easier to mount. Both are forms of persistence. Both show that "remembering" need not mean replaying a past event; it can mean being changed by the past in a way that affects future choices.

We should also notice what is gained and what is lost by this kind of memory. A slime trail is local and external; it helps avoid revisiting a patch of ground. An epigenetic shift is internal and systemic; it may prime a lineage of cells for heightened defence. Neither provides the rich, contentful recollection that humans associate with memory. But neither is trivial. Both are effective, and effectiveness is the currency that evolution counts.

If we zoom out, the immune system begins to look less like an army with generals and more like a city responding to an emergency. There are patrols, alarms, roadblocks, specialised units, and repair crews. Coordination emerges from communication networks: cytokines and chemokines as messages, gradients as signposts, adhesion as crowd control. There are also failures that resemble civic failures: overreaction, misdirected force, breakdowns in communication, chronic states of alert that become harmful in themselves.

A historical anecdote fits here, not as a single experiment but as a broader transformation in how we imagine immunity. For much of the twentieth century, the immune system was framed through metaphors of identity and discrimination: self versus non-self, invaders versus defenders. These metaphors captured something real, but they can mislead by implying that the immune system "recognises" in the way a mind recognises. The more we watch immune cells in vivo, the more the emphasis shifts from abstract recognition to dynamic organisation: who goes where, when, and how quickly; how signals are amplified and contained; how local interactions scale into tissue-level behaviour. The immune system is less a courtroom issuing verdicts than a crowd responding to smoke.

This brings us back to the book's central question: what does it mean to call such a system intelligent? If intelligence means conscious deliberation, then the immune swarm is not intelligent. If intelligence means the ability to solve problems by using information to control outcomes, then it is difficult to deny. The immune system solves the problem of localising damage, of allocating resources, of balancing aggression with restraint, and it does so with no central planner. It is a distributed control system that evolved to operate in a noisy, adversarial world.

There is one more reason the immune swarm matters for our broader argument. In bacteria and slime moulds, the agent and the environment are distinct: the cell moves in a world. In immunity, the "world" is you. The immune system is cognition turned inward, a problem-solving apparatus embedded within the body it must preserve. That makes its mistakes feel personal, because they are. Allergies, chronic inflammation, autoimmune disease: these are not failures of will or intelligence in the everyday sense, but failures of distributed decision rules, mis-tuned swarms, alarms that will not switch off.

And yet, when it works, it works so quietly that we forget it is happening at all. The splinter heals. The redness fades. The swarm disperses. The body returns to its ordinary state, as if nothing occurred. The intelligence of the immune system is, in this sense, an intelligence that aims to erase its own traces.

The next step in our story lies beyond the "cognitive speck" and even beyond the swarm. Cells can navigate, remember, and coordinate, but they can also build. Developmental biology—the process by which bodies form reliable shapes and repair themselves—requires problem-solving on the scale of tissues and organs, with electrical and chemical signals that operate over longer distances and longer times. If swarms show how intelligence emerges from crowds, development will show how intelligence emerges from *constraints*: how living matter maintains form, corrects errors, and pursues anatomical goals without ever drawing a blueprint.

Chapter 4

The Electric Blueprint

While DNA provides the molecular parts list for life, the instructions for assembly are written in a transient, crackling language of electricity. By eavesdropping on the whispers between cells, we reveal a hidden layer of information that determines whether a cluster of cells becomes an eye, a limb, or a tumor. This invisible scaffold exists before the physical body takes shape, guiding growth like a magnetic field arranging iron filings.

4.1 The Spark of Form

If you peel away the drama of nerves and muscles, electricity still hums quietly in the rest of the body. It is easy to miss because it rarely announces itself with motion. A neuron fires; a muscle contracts; a heart beats. Those are loud electrical stories. But beneath them sits a gentler and more persistent kind of electrical order: the resting voltage of ordinary cells, and the way many cells knit their voltages together into tissue-wide patterns. That distributed electrical pattern is not merely a by-product of being alive. In many contexts it behaves like a set of instructions that cells can consult, maintain, and sometimes rewrite.

A single cell is, electrically speaking, a tiny battery. Its membrane is an insulator separating two salty solutions: the cytoplasm inside and the extracellular fluid outside. Ions (charged atoms like sodium, potassium, chloride, and calcium) do not freely cross the membrane; they require proteins—ion channels and pumps—that act as selective gates and machines. Because different ions are allowed through at different rates, and because pumps actively move ions in particular directions, a charge difference builds across the membrane. This is the *resting membrane potential*, often abbreviated as V_{mem}. In a typical animal cell, the inside is negative relative to the outside, often tens of millivolts. That number sounds small, but across a membrane only a few nanometres thick it corresponds to a very steep electric field.

The common temptation is to treat V_{mem} as a mere housekeeping variable, the electrical equivalent of water pressure in plumbing. Water pressure matters, but it is not a message; it is just what happens when pipes and pumps exist. Yet biology has a habit of turning physics into language. Cells do not only endure voltage; they read it. Many proteins change their behaviour when the membrane voltage changes. Some channels are voltage-gated; some transporters speed up or slow down; some enzymes shift conformation; and gene expression can be pushed toward one fate rather than another. A voltage change can move more than charge. It can move *meaning*.

The most interesting twist comes when we stop looking at one cell at a time. In a tissue, cells sit side by side like apartments in a building. Between many neighbouring cells are structures called *gap junctions*: clusters of protein channels that directly connect the cytoplasm of one cell to the cytoplasm of the next. If the membrane is the wall, gap junctions are gated tunnels through it. They allow ions and small molecules to pass, which means that a voltage state can spread, smooth out, and form gradients. The electrical state becomes not just a property of a cell but a property of a community of cells—a network.

This is the point where electricity begins to resemble something we are more comfortable calling memory. In the brain, memories are not stored in single neurons but in patterns distributed across networks. The details differ, but the analogy is useful because it shifts attention from *components* to *collective states*. A tissue joined by gap junctions can settle into stable electrical configurations. Perturb it briefly and it may return to the same configuration, as if correcting an error. Push it past a threshold and it may settle into a different stable configuration. Such stability is the hallmark of memory-like behaviour: the present state depends not only on current inputs but on history.

One way to make this feel less abstract is to picture V_{mem} not as a single reading on a meter but as a map. Imagine laying a translucent sheet over a developing tissue and colouring each spot according to its voltage: deep blue for strongly negative (hyperpolarized), warm red for less negative or even positive (depolarized). The result is a heat-map of electrical state. In some embryos, those maps have striking structure long before any visible anatomy appears. A boundary in voltage can presage a boundary in anatomy. A patch of hyperpolarization can foreshadow where an organ will arise. The electrical pattern is not a photograph of the finished body, but it can function like an early sketch: constraints and cues that later biochemical and mechanical processes elaborate into flesh.

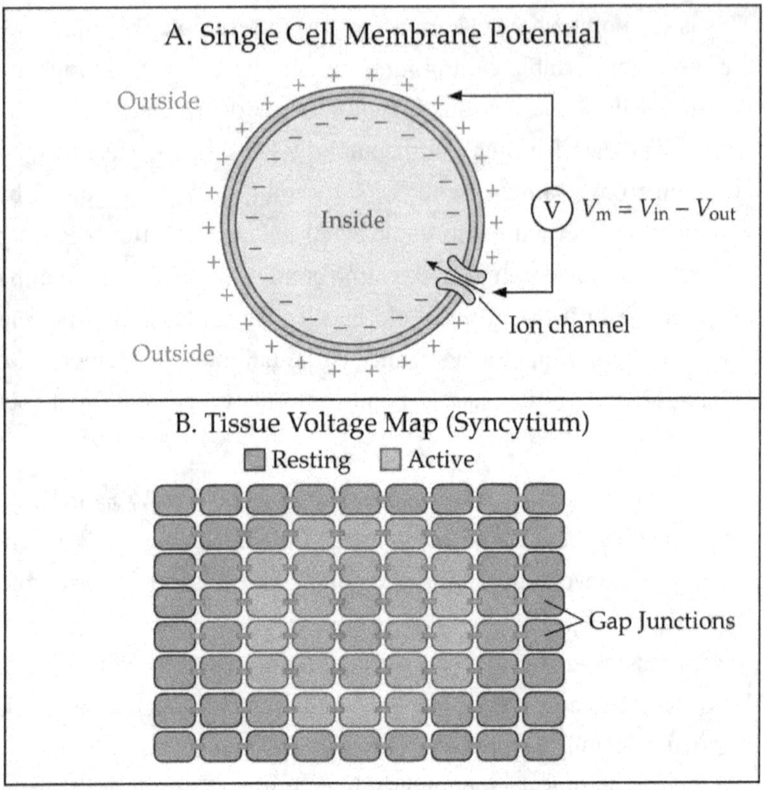

The idea that electrical properties might guide form has deep roots, though for much of modern biology it sat in the background. One of the earliest public moments came in the 1780s, when Luigi Galvani, an anatomist in Bologna, noticed that a dead frog's legs could twitch when touched by metal during dissection. The details of his apparatus mattered, and his interpretation was not fully correct, but his intuition was audacious: living tissue had its own electricity. Galvani's demonstrations were compelling enough to provoke a famous dispute with Alessandro Volta, who argued that the metals were creating electricity. Volta was right about the metals, yet Galvani was right about something more important: animals are not electrically inert.

That argument, played out with frogs, metal arcs, and improvised circuits, helped launch electrophysiology.

For a long time electrophysiology became, in practice, the science of excitable tissues. The spectacular spikes of neurons and muscles were easier to measure and easier to link to obvious outcomes. Meanwhile, the slow voltages of skin, epithelia, and embryos looked modest and ambiguous. It did not help that developmental biology, in the twentieth century, found its dominant language in genes and chemicals. If you could stain for a transcription factor or knock out a gene, you could tell a crisp causal story. Voltage seemed slippery: dynamic, distributed, and affected by many proteins at once. That slipperiness, however, is precisely what makes it a good medium for higher-level control.

To see why, consider how a tissue can use voltage as a shared state. Each cell has channels and pumps that tend to push its membrane potential toward some value. But because cells are coupled through gap junctions, no cell is an island. A depolarized patch can influence its neighbours; a hyperpolarized region can stabilize itself by recruiting current from surrounding areas. Over time, the tissue can form domains: regions with distinct and persistent electrical character. Those domains can be read by downstream pathways. A cell sitting on one side of a boundary may experience a different balance of ion flux, pH, calcium dynamics, and small signalling molecules than a cell on the other side. The boundary is not a line drawn in ink; it is a maintained difference in physiological state.

What does it mean to say that such a pattern stores a memory? Not memory in the sense of recalling a childhood holiday, of course. The memory here is about *target morphology*—a tendency to build and rebuild a particular anatomy. The most familiar example is regeneration. If you cut your skin, it heals into skin, not into bone or a random lump. If a salamander loses a limb, it can regrow a limb with the right parts in the right places. Even animals with limited regeneration exhibit a stubbornness of form: wounds close, tissues reorganize, and growth usually stops when the right size is reached. The system behaves as if it has a set-point: a preferred large-scale outcome.

Genes are indispensable to this story, but they are not sufficient to make it comprehensible. DNA is a catalogue of parts and potential behaviours. The puzzle is how those parts are coordinated in space and time to achieve a coherent, scaled body plan, and how the plan is maintained in the face of disturbance. Here the electrical network offers a different kind of explanatory foothold. Because V_{mem} is a state variable that can be shared across cells, it can represent something like a collective decision: not just what a single cell should become, but what the tissue should be doing as a whole.

A concrete example comes from early vertebrate development, where researchers have observed characteristic voltage patterns in embryos before organs form. The striking point is not merely that voltage correlates with later anatomy; it is that manipulating voltage can change anatomical outcomes. If you alter the activity of certain ion pumps or channels in the right place and time, you can disrupt the normal formation of structures; if you impose a different bioelectric state, you can sometimes coax tissues into building structures in unusual locations. The implication is not that electricity replaces genetics, but that bioelectric patterning is a layer of control that can sit upstream of familiar genetic programs, steering when and where they are deployed.

Gap junctions make this steering inherently non-local. A cell does not need to contain all the information required to behave appropriately; it can be guided by the electrical context created by its neighbours. This helps explain a feature of development that otherwise feels magical: robustness. Many embryos can tolerate missing cells, shifted cell positions, or mild perturbations and still arrive at a normal form. Robustness is what you expect from a system that has feedback: deviations are detected and corrected. Electrical coupling is a natural substrate for such feedback. A tissue-wide voltage pattern can act like a reference signal. If one region drifts, coupling and channel dynamics can pull it back, or trigger compensatory responses.

There is a further, subtle reason voltage patterns can be memory-like: many ion channel and pump activities are regulated *post-translationally*. That is, the same genetic toolkit can produce different electrical states depending on how proteins are gated, trafficked, and modulated by the cell's internal chemistry. Two tissues may express similar sets of channels yet sit in different stable electrical configurations, just as two computers with identical hardware can run different software states. This makes V_{mem} a powerful integration point. It compresses many molecular details into a single, actionable variable that other pathways can read.

If this sounds like an attempt to smuggle cognition into places it does not belong, it helps to be precise about what is being claimed. No one needs to imagine that a limb bud is conscious, or that an epithelium has feelings about its voltage. The word *intelligence* in this book is being used in the sober sense: the ability to achieve goals across changing conditions. From that angle, a tissue that maintains a target anatomical pattern, corrects mistakes, and adapts to injury is showing a kind of problem-solving competence. Bioelectric networks offer a plausible physical basis for that competence, because they can represent global states, propagate information, and support feedback.

Once you begin to view a tissue's electrical state as a pattern, new questions present themselves almost immediately. Where does the pattern come from in the first place? How does a linear genome, copied into every cell, give rise to a three-dimensional organism with stable proportions and reliable landmarks? And why does the body so often behave as if it is following a plan that exists *before* the parts are built? The next step is to confront a surprisingly stubborn mystery: living systems routinely build complex structures without anything that looks like a conventional blueprint, yet the result is specific enough that we recognize a face, a hand, a jaw. The electrical sketch is only the beginning of that story, but it hints at how biology can turn physics into a pre-pattern—a quiet scaffold on which genes and cells later hang the visible form.

4.2 Building Without Blueprints

A genome is often described as a blueprint. The metaphor is flattering to DNA, and reassuring to us, because blueprints belong to a familiar world: a plan on paper, a builder on site, and a finished structure that matches the drawing. But embryos do not read drawings. They do not have an architect standing outside the system, checking measurements and issuing corrections. An embryo is a crowded, wet, noisy place where cells divide, push, pull, and migrate while chemical signals diffuse and decay. If anything deserves the name "blueprint" in such a setting, it cannot be a static script alone. It has to be something that can *run*.

The previous section sketched one candidate for such a running plan: tissue-wide patterns of membrane voltage, maintained and shared through gap junctions. That layer of organisation is not a substitute for genes. It is better thought of as an interface between genes and geometry, between the molecular parts list and the large-scale problem of making a coherent body. To see why, it helps to state the developmental puzzle in its bluntest form.

Every cell in your body (with a few exceptions such as red blood cells) carries essentially the same DNA. Yet cells in different places do different things, and they do them at the right times, and in the right proportions. A liver grows to fit the body; a finger stops at a finger. Wounds close; organs remain separate; left and right generally agree on what counts as left and right. Developmental biologists sometimes call this an "inverse problem": we can read the genome as a linear sequence of letters, but we are asked to explain how that sequence yields a three-dimensional, self-repairing form. The inverse problem is difficult not because genes are unimportant, but because genes are *local*. Each cell can only transcribe its own DNA, sense its own environment, and talk to its neighbours. The body, meanwhile, is a global outcome.

Static code can produce a global outcome if the system has a way to generate and maintain patterns across space. Chemistry offers one route, through gradients and morphogens: substances that diffuse and instruct cells according to concentration. Mechanics offers another, through tension, stiffness, and the physical constraints of tissue. Bioelectricity adds a third: patterns of voltage and connectivity that spread through networks and can persist as stable states. The key word is *state*. A state is not just what parts you have; it is how those parts are currently configured, and what they are therefore likely to do next.

The difference matters in a practical way. If you look at a piece of tissue and list the genes it expresses, you have not necessarily captured the tissue's control variables. Many of the proteins that set membrane voltage are regulated after they are made. An ion channel can be present but closed. A pump can be moved to the membrane or tucked away inside. A gap junction can be open, partially open, or sealed off. Two tissues with similar gene expression can therefore occupy different electrical states, just as two identical radios can either play music or sit silent depending on how their dials are set. The genome supplies the components; physiology supplies the configuration.

Here is a simple thought experiment that makes this concrete. Suppose an embryo needs to establish a boundary: these cells will become part of the face, those will become part of the braincase. One way is to use a chemical gradient and have cells read concentration thresholds. Another is to use a voltage boundary: a region of hyperpolarization here, depolarization there, with a sharp transition maintained by patterns of ion channel activity and by which cells are electrically coupled. The chemical method is powerful but vulnerable to the physics of diffusion and degradation; it is also slow to correct if something goes wrong. The voltage method is not magical, but it has a useful feature: because current can flow quickly and because coupled tissues tend to smooth irregularities, electrical boundaries can be robust. They can behave like maintained lines, not just fading clouds.

The moment you permit such state variables, the metaphor of DNA as blueprint begins to look incomplete. A blueprint is static: it does not change when the building wobbles in an earthquake. A functioning control system is different. It senses deviations, compares them to a target, and acts to reduce the mismatch. In engineering language it is feedback; in everyday language it is error correction. In biological development and regeneration, error correction is not a luxury. It is the reason a body does not drift into monstrosity every time a few cells die, or a limb is bruised, or a growth signal arrives a little late.

This brings us to a historical figure who, long before anyone could measure embryonic voltages, understood the necessity of self-organised patterning: Alan Turing. Turing is best known for his wartime codebreaking and his foundational work in computer science, but in 1952 he published a paper with a strange, almost domestic title: *The chemical basis of morphogenesis*. The paper asked how simple chemical interactions could produce the recurring patterns we see in nature: stripes, spots, spirals, and repeated structures. Turing proposed what is now called a reaction—diffusion system: two (or more) substances that react with each other and diffuse through a tissue at different rates. Under the right conditions, a uniform field becomes unstable and breaks into a pattern. No external blueprint is needed; the pattern emerges from the dynamics.

What made Turing's idea so daring was that it treated development not as an execution of a plan but as the unfolding of a computation. The tissue, in effect, solves a problem: how to turn a nearly homogeneous starting state into an organised spatial arrangement. Turing's mathematical patterns were chemical, but the broader lesson is not about chemistry. It is that pattern can be an attractor of a dynamical system, and attractors can serve as the "targets" toward which development moves.

Bioelectric pre-patterns fit naturally into that way of thinking. A gap-junction-coupled tissue is a dynamical system. Each cell's membrane voltage depends on ion channels, pumps, and transporters; the coupling depends on the conductance of junctions between cells. Together they define a set of possible electrical configurations. Some configurations are unstable: a small disturbance sends the system elsewhere. Others are stable: the tissue returns to them when perturbed. Those stable configurations are attractors. If a particular attractor corresponds to "build an eye here" or "stop growth now," then the electrical state functions as a kind of memory of the target morphology. It is a memory not stored in a molecule like DNA, but in a distributed pattern that the tissue can maintain.

The phrase "pre-pattern" deserves emphasis because it highlights a temporal ordering that can otherwise be missed. In many developmental systems, an electrical pattern appears *before* the visible anatomical structure, and in some cases before the familiar gene expression markers of that structure become obvious. This is not mystical foresight; it is causality in the direction that control systems often take. If a later event depends on a global context, the global context must be established early, when the system is still plastic enough to adopt large-scale organisation.

Consider a generic organ-to-be: a patch of tissue that will later differentiate into something recognisable. Long before the organ has cells that look special under a microscope, the future region has to be singled out. Some of that singling-out comes from chemical signals, some from mechanical constraints, and some, evidence suggests, from physiological state. The early electrical pattern can be thought of as a set of biases: boundaries where coupling is stronger or weaker, regions where voltage makes certain gene networks easier to activate, and paths along which signals can travel. It is a sketch laid down in voltage, later filled in with cell types, extracellular matrix, blood vessels, and nerves.

One reason this sketch is so useful is that it can unify diverse genetic mechanisms. If you knock down one ion channel, another can sometimes compensate, not necessarily by matching the channel's molecular identity but by restoring the *physiological outcome*. This is a subtle but profound form of redundancy. It is not redundancy of parts; it is redundancy of states. Multiple molecular routes can reach the same V_{mem} pattern. In such a situation, the effective control variable is not "channel A" but "voltage domain X." That helps explain why some genetic manipulations have surprisingly mild phenotypes: the system's goal may be encoded at the level of a tissue state, and the system can reach that goal by alternative molecular paths.

If this begins to sound like special pleading—a way to excuse any complexity by invoking a higher-level pattern—it is worth anchoring the idea in an everyday analogy. Think of a thermostat. The blueprint of the thermostat (its wiring diagram) matters, but it does not tell you the temperature of the room. The temperature is a state variable produced by the thermostat's interaction with the heater, the insulation, the weather, and the habits of the occupants. If you want to predict whether the room will be comfortable, you cannot read the wiring diagram alone. You have to know the state and the feedback rules. In development, genes are like the wiring diagram; bioelectric patterns are closer to the current settings and the internal signals that keep the system near its desired outcome.

Bioelectricity has another feature that makes it especially suited to linking local and global information: it is inherently relational. A voltage is a difference between inside and outside. A voltage gradient is a difference between one region and another. When cells are coupled, each cell's voltage is partly a function of its neighbours' voltages. That relational character makes it easier to represent shape, because shape is also relational. A boundary is not a thing; it is a difference. Proportion is not a gene; it is a relationship between sizes. Symmetry is not a protein; it is a constraint connecting one side of the body to the other.

There is also a cultural history here, a kind of collective blind spot in modern biology. The twentieth century rewarded what was easy to catalogue: genes, proteins, signalling pathways. These are discrete objects. Bioelectric patterns, by contrast, are fields and networks; they require measurement strategies that many laboratories simply did not have. It is telling that, when early embryologists spoke of "fields" and "organisers," they were often accused of hand-waving. The accusation was not entirely fair. They were reaching for a concept that our tools could not yet make concrete. As fluorescent voltage-sensitive dyes, genetically encoded reporters, and optogenetic actuators became more common, the field-like language began to regain respectability, not as mysticism but as a description of measurable state.

The difference between a static blueprint and a dynamic pre-pattern becomes clearest when we ask how development deals with surprises. Suppose a cluster of cells that would normally contribute to a structure is removed. Many embryos can compensate: neighbouring cells change fate, boundaries shift, and the final outcome is closer to normal than the perturbation would suggest. Such compensation is hard to explain if the system is executing a rigid script. It is easier if the system is aiming for a stable attractor: a pattern that can be reached from multiple starting points. The electrical network, coupled through gap junctions, is a plausible substrate for that aiming because it can quickly integrate local changes into a tissue-wide state and, through downstream pathways, adjust cell behaviour accordingly.

This is where the term "intelligence" becomes tempting again, and again needs careful handling. There is a kind of intelligence that is simply control: the ability to steer a system toward a goal in the face of noise and disturbance. A thermostat has that kind of intelligence without awareness. So does a bacterial chemotaxis network. So does a tissue that corrects its growth trajectory. The genome cannot provide that intelligence as a static document; it can only provide the machinery. The intelligence resides in the dynamics, in the way the machinery is wired and how its current state constrains what happens next.

At this point a reader might reasonably ask: if the electrical pattern is so important, why do we not see it in standard molecular data? Why does a transcriptome not reveal it? The answer is that transcripts are like inventory lists. They tell you what parts are present, but not how they are being used at this moment. Voltage depends on conductances, gradients, and couplings; it can change without changes in mRNA. A tissue could quietly shift its physiological set-point with minimal transcriptional drama, and the consequences would only become visible later, when gene networks respond, when cells migrate differently, or when proliferation changes. The causal arrow can point from state to gene expression as often as the other way around.

The idea of pre-patterns also softens a common misunderstanding about genes and destiny. If DNA were truly a blueprint in the architectural sense, then knowing the genome would be close to knowing the form. In reality, knowing the genome is more like knowing the alphabet and grammar of a language without knowing what sentence will be spoken. The sentence depends on context, history, and interaction. Bioelectric patterns are part of that context: a tissue-level "accent" that shapes which molecular sentences are likely to be uttered.

One of the most provocative implications is that there may be multiple stable "sentences" available to the same genetic system. The same genome can, under different physiological states, settle into different morphological outcomes. Some outcomes may be rare in normal development, suppressed by the usual attractor landscape, yet reachable if the system is pushed. This is not speculation in the abstract. Regenerative systems, especially those that can rebuild complex anatomy after drastic perturbation, are the natural arena where such alternative attractors can be revealed. When an animal regenerates, it is not merely repairing; it is re-solving the inverse problem under unusual boundary conditions.

The next section takes advantage of that arena. If bioelectric patterns are truly part of a dynamic blueprint, then changing the pattern should change the build. Not by rewriting DNA, but by nudging the tissue into a different stable state, as if changing the settings on a self-assembling machine. Regeneration, with its dramatic before-and-after clarity, offers some of the most unsettling tests of this idea. When researchers deliberately perturb voltage gradients and gap junction connectivity, tissues sometimes do not merely fail; they choose a different anatomical destination. That is where the metaphor of an "electric blueprint" stops being poetic and starts to feel like a hackable control system, and that is precisely the territory we now enter.

4.3 Hacking the Growth Code

There is a special kind of astonishment reserved for regeneration. A cut closes; a scar forms; we nod and move on. But watch a creature rebuild a missing part with the right geometry, the right orientation, and the right stopping point, and the familiar story—genes make proteins, proteins make tissues—begins to feel incomplete. Regeneration looks less like construction and more like remembering. The animal behaves as if it knows what is supposed to be there.

If the earlier sections did their job, the reader now carries a quiet suspicion: perhaps the body has a control layer that is not written as DNA sequence but as a living, adjustable state. Bioelectric patterns—distributions of membrane voltage shared through gap junctions—are plausible candidates for such a layer. But plausibility is not proof. Biology is full of correlates that dissolve under experimental pressure. The decisive question is whether bioelectric state can be *rewritten* to change anatomy, while leaving the genome untouched.

To answer that, we need experiments that do two things at once. They must show that a bioelectric intervention happens early enough to be a cause, not a downstream echo. And they must show that the anatomical outcome is not a vague sickness phenotype—not simply slower growth, more cell death, or general toxicity—but a *patterning* change: a different structure, in a different place, or with a different identity. Few systems are as bluntly revealing as planarian flatworms.

Planaria are small, soft-bodied freshwater animals that have become minor celebrities in laboratories because they regenerate with unnerving confidence. Cut one in half and each half will regrow what is missing: head from tail, tail from head. Cut it into many pieces and many pieces will build whole worms. This is not simply rapid cell division. The animal must decide which end becomes a head, where to put eyes, how long to make the body, and when to stop. For decades, researchers asked which genes and chemical gradients guide those decisions. More recently, experiments from Michael Levin's lab, including Durant and colleagues in 2017, have shown that electrical connectivity and voltage states are not just helpers but high-level control knobs.

One of the most provocative results is the induction of two-headed worms, reported in Durant's 2017 work with the Levin group. The headline is easy to misunderstand, so it is worth being careful about what is meant. The point is not that a scientist "made a monster" by crudely damaging tissue. The point is that a transient manipulation of bioelectric signalling can cause a regenerating fragment to adopt a different *target morphology*—a different stable outcome—and then maintain that outcome as if it were normal. The genome has not been edited. The animal has not been permanently poisoned. The patterning memory has been shifted.

A simplified version of the logic goes like this. In a normal regenerating fragment, cells at the wound site must choose a head programme or a tail programme. Many biochemical pathways participate, but the global decision is surprisingly sensitive to the electrical state of the tissue network. Gap junctions, the tunnels that couple neighbouring cells, are one way to tune that network. If you block gap junction communication during the critical early window after amputation, the tissue no longer behaves as a single coordinated electrical unit. Under certain conditions, this disruption allows alternative stable outcomes to appear. Some fragments regenerate as two-headed worms.

This is not merely a matter of "more head tissue." A head has organised structures: a brain-like ganglion, eyespots, a mouth positioned along the body axis. Two-headed regeneration shows a coherent re-specification of the body plan. It is as if the system that normally enforces the rule "one head, one tail" is a control circuit, and the circuit has been pushed into a different stable state.

What makes the story even more unsettling is that, in some experiments, the effect can persist beyond the initial insult. A worm induced to regenerate with two heads can later be cut again and, rather than reverting to the usual one-head pattern, regenerate as two-headed *again* even without further treatment. That persistence is hard to reconcile with a purely local wound response. It looks like a rewritten set-point, a memory stored in the physiological state of the tissue.

The language of dynamical systems helps here. A tissue network can have multiple attractors: stable patterns of activity toward which it tends to settle. In a neural network, such attractors can correspond to memories or categories. In a regenerative tissue, attractors can correspond to morphological outcomes. A brief perturbation can push the system from one basin of attraction into another. Once there, the system's own feedback maintains the new state. Under this view, two-headed regeneration is not a freak accident but a stable alternative solution that the system normally avoids.

A clean way to depict the logic is to show the causal chain visually: a change in voltage-map and connectivity, followed by a change in anatomical identity. The point is not that we know every intermediate step, but that the intervention has a coherent and legible direction: alter bioelectric state, and the pattern of regeneration shifts accordingly.

Bioelectric State Rewrites Outcome (Planaria Example)

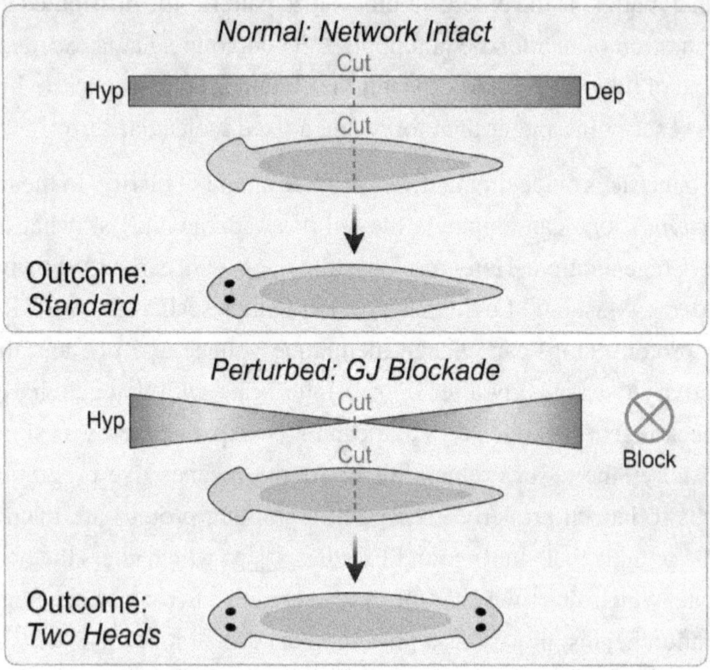

Physiological state overrides genetic default

A skeptic could still object: perhaps blocking gap junctions is simply a kind of developmental vandalism. If you interfere with cell–cell communication, odd things happen; why treat the odd outcome as evidence of a stored pattern? The answer lies in the specificity and the persistence. A mere injury tends to produce disorganisation. Here, the outcome is often organised, repeating, and stable. The tissue is not failing to regenerate; it is regenerating according to a different rule.

The planarian story also makes a second point that is easy to miss. The control variable is not the identity of a particular gene. Many different molecular manipulations can shift the tissue's voltage and connectivity into comparable physiological states. In such cases, what matters is the emergent bioelectric pattern: the distributed configuration of membrane potentials and coupling. That is exactly the kind of thing one would expect if the system is operating with a high-level set-point, rather than following a fixed molecular script.

If planaria provide drama, frogs provide a kind of clarity. In the frog *Xenopus*, one can amputate the tail of a tadpole and ask what triggers regeneration. Here, too, physiology appears early, as shown by Adams, Masi, and Levin in 2007. Ion pumps such as the V-ATPase (a proton pump) can change membrane voltage and pH, and their activity rises quickly after injury. Interfering with that activity can block regeneration in ways that cannot be explained away as simple toxicity: the tissues remain alive, yet the regenerative programme fails to launch properly. The point is not that protons are magical. The point is that altering ion flux alters V_{mem}, which alters the tissue state, which alters whether the system commits to rebuilding. Regeneration begins, in part, as a physiological decision.

The most exciting twist in this frog work is that bioelectric state can be manipulated not only with drugs or genetic knockdowns but with light, as Adams, Tseng, and Levin showed in 2013. Optogenetics is usually introduced as a neuroscience technique: insert a light-sensitive protein into neurons and use flashes of light to activate or silence them. But the logic applies to any cell that maintains a membrane potential. If you express a light-driven proton pump such as Archaerhodopsin in non-neural tissues, shining light can hyperpolarize those cells. This is a rare gift to experimental biology: a way to *write* a physiological state with temporal precision.

The temporal detail matters because development and regeneration are full of windows. There are moments when a decision is being made and later moments when that decision has already been committed to tissue architecture. In *Xenopus*, experiments suggest that there is an early period after amputation during which forcing a particular bioelectric state can rescue regenerative capacity even when it would normally be reduced. A couple of days of intervention can have consequences that last weeks, because it nudges the system back into the correct attractor basin early enough for downstream gene networks, innervation, and growth dynamics to unfold.

At a human level, this kind of result invites both hope and caution. Hope, because it suggests a new route toward regenerative medicine: instead of trying to micromanage every growth factor and scaffold, perhaps we could restore the body's own control settings. Caution, because ion channels and pumps are everywhere. The same proteins that set membrane voltage in a wound also operate in the heart, the brain, and the gut. If bioelectricity is a control layer, then it is a deeply shared control layer. Hacking it carelessly would be like adjusting the voltage of an entire city to fix the lighting in one street.

This brings us to a darker mirror of regeneration: cancer. Cancer is often described as cells that grow too much. That is true, but insufficient. Cancer cells also forget the social rules of tissues. They invade, migrate, ignore boundaries, recruit blood supply, and corrupt their neighbourhood. In other words, cancer is not just excess growth; it is loss of pattern control. It is tempting, therefore, to ask whether bioelectric set-points that guide normal patterning also participate in keeping cells behaving as parts of a coherent whole.

A growing body of work links depolarized resting potentials and disrupted tissue electric fields to behaviours associated with invasion and metastasis. The details are complicated and the mechanisms are still being mapped, but the broad idea is compelling: if a tissue's electrical state helps enforce "you are liver, stay in liver, stop when the liver is the right size," then disturbing that state could loosen the constraints that make a tissue a tissue. Conversely, restoring more normal bioelectric patterns might push cells back toward cooperative behaviour, not by killing them but by reasserting the collective rules.

This is not a promise of an electric cure for cancer. It is a reframing that could influence how we search for therapies. Standard approaches often aim to block a specific mutation or pathway. But if the effective control variable is a physiological state, then many different molecular disruptions might converge on similar electrical outcomes. In that world, it could be less important to block *this* oncogene than to restore *this* tissue-level set-point. That is a radical shift in strategy, and it comes with formidable obstacles: measurement in living human tissues, specificity of intervention, and the possibility of unintended patterning changes. Yet the concept is too powerful to ignore, because it speaks to the central mystery of multicellularity: how trillions of cells agree on a single body.

A useful way to hold all of this in mind is to distinguish between *parts* and *policies*. Genes and proteins are parts. Bioelectric states, coupled across tissues, look more like policies: rules and set-points that coordinate parts toward a goal. When you change a part, you may get a local defect. When you change a policy, you may get a different global outcome.

There is also a philosophical sting here. We are used to thinking that the genome is the deepest layer of biological identity. It feels like essence: change the DNA, change the creature. The bioelectric experiments complicate that picture. They suggest that an organism's form is not specified once, at fertilisation, but continually negotiated by a living network. The genome builds the network, but the network, through its state, stores something like a memory of what the body should be. That memory can be edited.

If that is true, a final question becomes unavoidable: how do we read and write these electrical patterns with the same ease that modern biology reads and writes genes? The tools are emerging—

voltage-sensitive reporters, optogenetic actuators, pharmacological modulators—but a tool is not yet a theory. To use bioelectricity as a genuine control panel for anatomy, we would need maps of the relevant state spaces, an understanding of which perturbations move the system between attractors, and a way to predict the downstream consequences. In other words, we would need a science of *pattern control.*

The chapter so far has argued that tissues can behave like problem-solving systems without brains: they store target states, correct errors, and sometimes switch to alternative stable outcomes. The next step is to treat this seriously as a kind of cognition—not conscious, but competent. What does it mean for a tissue to have a goal? How is that goal represented, and how far can it be pushed before the system breaks? The experiments that give us two-headed worms and rescued frog tails are not merely curiosities. They are invitations to rethink what biological intelligence looks like when it is spread across cells, written in voltage, and expressed as form.

Chapter 5

The Worm That Never Forgets

Slice a planarian into nearly three hundred pieces, and in a few weeks, you will have three hundred perfect worms. This chapter examines how this humble creature maintains a memory of its own anatomy not in its genes, but in a modifiable bioelectric network. By observing how these worms can be reprogrammed to build heads where tails belong, we discover that the body's shape is a computational goal state that tissues actively strive to achieve.

5.1 The Immortal Blueprint

A planarian looks like something you might scrape off a rock without noticing: a few centimetres of soft, flattened body, two dark eyespots, and a slow, gliding motion that seems more like drifting than walking. Yet this small animal has become one of biology's most unsettling characters, because it refuses to behave like an individual object. An earthworm is a body: damage it badly enough and it dies. A planarian is closer to a process: interrupt it, divide it, and it reorganizes itself back into something whole.

The headline fact is easy to say and hard to internalize. Cut a planarian into pieces, and many of those pieces will grow into complete worms. Not *survive* as wounded remnants. Not *heal* in the ordinary sense. They rebuild missing organs, restore proportions, and re-establish the head-to-tail axis with a reliability that feels almost intentional. The animal's apparent individuality is, in a strange way, negotiable: one becomes two, two become three, and each result is recognizably a planarian.

That is the spectacle. The deeper puzzle sits behind it, and it is not mainly about fast growth or prolific stem cells (though those matter). The real question is informational. A fragment has to do something more sophisticated than merely produce tissue. It has to decide *what kind* of tissue to produce, *where* to place it, and *when* to stop. A severed tail piece must choose to grow a head at one end and not at the other. A trunk fragment must produce both a head and a tail. A sliver must scale its new anatomy so that the regenerated worm is not a head the size of a pin on a body the length of a pencil, but a coherent miniature.

This is the *survey problem*: how does a piece of living material, suddenly deprived of the rest of its body, figure out what it is missing? Somehow it must take stock of its current state relative to a target. It is as if every part carries an internal blueprint of the whole, and can compare "what I have" to "what I should be."

The word *blueprint* is a dangerous metaphor, because it suggests a static diagram that a builder simply follows. Living tissue does not work like a carpenter reading a plan. Cells do not unroll a scroll. They communicate locally, they respond to signals, they change their behaviour based on neighbours and conditions. Still, the metaphor helps because it points to something undeniably real: a planarian fragment behaves as if a global pattern is available to it, even when the global body is gone.

To appreciate why this is so provocative, it helps to remember what most of us were taught, implicitly if not explicitly, about development. We imagine that the genome is a kind of recipe, and development is the act of cooking. Follow the recipe and you get an organism. Damage the organism and you may be able to patch it, but you cannot, in general, cook the same dish from a spoonful scraped from the plate. The planarian mocks that intuition. A fragment does not merely continue the recipe from where it left off; it recalculates the entire dish.

The historical fascination with planarians is older than modern genetics and even older than cell theory's mature form. In the eighteenth century, naturalists were enthralled by creatures that seemed to blur the border between injury and reproduction. Reports circulated of worms that could be divided and would still move, feed, and reform. These stories landed in the same intellectual space as hydra, starfish, and salamanders: living beings that made the question of "what counts as a whole animal" feel suddenly negotiable. The fascination was not only technical; it was philosophical. If an animal can be cut up and persist as many animals, what exactly has been divided? Matter? Identity? Agency?

By the time laboratory biology matured in the nineteenth and twentieth centuries, planarians had become reliable experimental partners. They were easy to keep, cheap to feed, and robust under manipulation. But what made them scientifically precious was not their convenience; it was their defiance of the usual trade-offs. Many animals can heal, and some can regenerate particular structures, but few can do it with such completeness and flexibility. In planarians, regeneration is not a special trick reserved for one organ; it is a general mode of existence.

The basic mechanics begin with an injury response. After amputation, a planarian seals the wound. Cells near the cut change behaviour, and a small mass of dividing cells forms at the site: a *blastema*, a bud of new tissue that will become whatever structure is missing. Beneath that visible bud lies something less visible and more important: a reorganization of information across the fragment. The fragment has to establish a new internal coordinate system: which end will be anterior (head) and which posterior (tail), where the midline lies, and how to scale organs appropriately.

None of that is obvious from the fragment alone. Consider a tail piece. It has only posterior tissues, no brain, no mouth, no eyes. Yet one wound will become a head. The other wound will remain tailward, rebuilding what was lost in cutting but not inventing a second head. If you had never seen this and were asked to predict what a tail fragment would do, you might guess it would simply close both ends and remain a tail. Or you might guess it would grow two heads, one on each cut surface, since each wound is "missing" anterior structures. Or perhaps it would grow something ambiguous, a malformed compromise. Instead, it typically grows one head in the right place, as if it knows which end used to face forward in the original animal and insists on restoring that direction.

The insistence becomes even more dramatic when fragments become small. Experiments have shown regeneration from surprisingly tiny pieces. Numbers are often repeated in classrooms and popular accounts for good reason: they sound like folklore, until you see the images. A fragment can be a small fraction of the original animal and still produce a complete planarian. One famous figure that circulates in the literature is a fragment on the order of *one two-hundred-and-seventy-ninth* of the worm, still capable of rebuilding a whole body. The exact fraction is less important than what it represents: below a certain size, almost any intuitive idea of "the part contains a miniature of the whole" breaks down. A tiny piece cannot literally contain a pre-formed head region or an obvious gradient of organs. Whatever guides regeneration must be encoded in a distributed way, accessible even when the available tissue is minimal.

This is where planarians force us to separate two kinds of explanation that are often conflated. One explanation says: planarians regenerate because they have the right *parts*—in particular, a large population of pluripotent stem cells called *neoblasts*, capable of generating many cell types. That is true, and it is a major reason planarians can rebuild organs. But it does not answer the blueprint question, because stem cells alone do not decide the architecture. A bag of bricks does not explain the design of a house. The other explanation says: planarians regenerate because they have a system of *positional information*—signals that tell cells where they are along the body axes and what they should become. This is closer to the heart of the matter, but it immediately raises a second question: positional information relative to what, exactly, when the body has been cut?

One way to grasp the strangeness is to imagine an everyday repair problem. Suppose you have a torn map of a city. If you have the whole map, you can see what is missing and draw it in. If you only have a small scrap—a few streets and a river bend—you cannot reconstruct the entire city unless you possess, in addition to the scrap, some *model* of the whole city. The planarian fragment behaves as if it has such a model. Yet it does not seem to store it in a single privileged spot, because cutting that spot away does not abolish regeneration. The model must be distributed, more like a pattern stored across a network than a document kept in a cabinet.

That network-like nature becomes clearer when one asks what regeneration must coordinate. It is not enough to regrow *a* head; it must be a planarian head, with species-specific shape, correct orientation, and appropriate scale. It must connect to the nerve cords, integrate with muscle and epidermis, and coordinate with the rest of the body so that feeding and locomotion work. A trunk fragment must decide where to place the pharynx, a muscular feeding organ, and how large it should be relative to the overall body. The new tissues must not remain as a raw overgrowth; they must be remodelled until the animal's geometry makes sense.

If this begins to sound like a kind of intelligence, that is not an accident. Intelligence, in the broad sense we are pursuing in this book, is not the presence of a brain or the ability to solve puzzles on command. It is the capacity to steer toward a goal despite disturbances. In a planarian, the goal is anatomical: a particular body plan. Injury is a disturbance. The animal's response is not a simple reflex but a sustained sequence of actions—cell division, migration, differentiation, and sculpting—coordinated over days. There is something here that resembles problem-solving: not conscious, not deliberate in the human sense, but organized in a way that achieves a robust outcome.

A useful word for this is *morphological homeostasis*: the maintenance of form. We are familiar with homeostasis in physiology. Body temperature is regulated around a target; blood sugar is kept within bounds. When the system is perturbed, feedback mechanisms act to restore the set point. Planarians hint that anatomy itself can be regulated in an analogous way. The fragment does not merely build once; it continues to adjust until the morphology matches a stored target. That target is what we mean by an *immortal blueprint*: not an indestructible drawing, but a durable, distributed memory of what the animal ought to be.

This way of framing it also clarifies why regeneration is not simply growth. Growth is easy to misunderstand as "more." Regeneration is "right." A tumour grows; it does not regenerate. A healing scar seals; it does not rebuild a missing hand. To regenerate is to keep track of a goal state and to deploy the right changes to reach it. The surprise in planarians is that this goal state seems to persist even when the organism has been reduced to a fraction of itself.

Some readers may already be objecting: surely the blueprint is just the genome. If every cell contains the same DNA, then every fragment contains the same instructions. But DNA is not a blueprint of the body in the way a diagram is a blueprint of a machine. DNA encodes proteins and regulatory elements; it provides the parts list and many of the rules of interaction. It does not, by itself, specify the current state of the organism or the dynamic choices that must be made after injury. Two cells with identical genomes can behave differently depending on context. A planarian tail fragment and a planarian head fragment contain the same DNA, yet one must build a head and the other must build a tail. The difference is not genetic; it is informational in a broader, physiological sense.

To see why, it helps to distinguish between *capacity* and *instruction*. Neoblasts provide capacity: the ability to generate cell types. Positional control systems provide instruction: a coordinate framework that tells those capacities what to do. But even positional information has to be updated after amputation, because the old coordinates referred to a larger body. A tail fragment cannot simply continue to believe it is at "90% posterior," because in a new worm it will eventually contain regions that are anterior relative to the new whole. Somewhere in the tissue, a mapping has to be recalculated: a new "where am I?" that is consistent with the new scale.

At this point we reach a fork in our intuitions. One intuition, inherited from classic developmental biology, says that chemical gradients and gene expression domains are enough: molecular signals set up by the wound determine head versus tail, and then downstream genetic programmes build the appropriate structures. Another intuition, less familiar but increasingly supported by experimental work, suggests that physiology itself—the electrical and communicative state of cells, coupled through tissues—provides a higher-level pattern memory that guides and constrains the molecular programmes. The planarian is not merely reading genetic instructions; it is computing an anatomical decision.

Even without getting ahead of ourselves, we can already locate the conceptual leap. A fragment is not simply missing matter; it is missing *relations*. A head is not a lump of cells, but a structured set of connections: nerves wired to muscles, sensory organs aligned to behaviour, mouth connected to gut. Regeneration requires reinstating these relations in the right geometry. That is a problem that looks much more like control and coordination than like construction from a parts list.

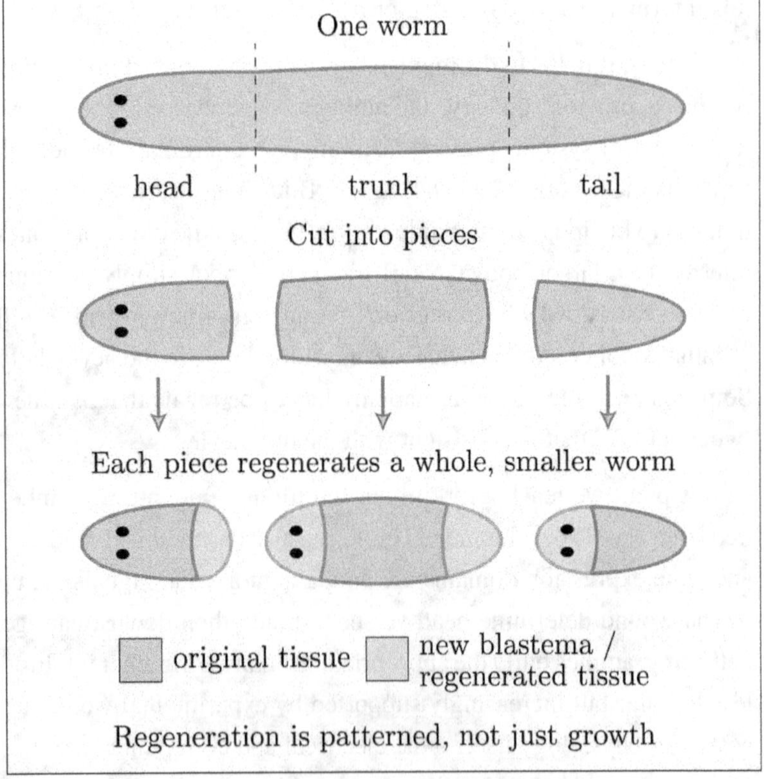

Regeneration is patterned, not just growth

The diagram hides as much as it reveals, because the most astonishing step happens before any new anatomy is visible. Within hours, long before a recognizable head bud forms, the fragment commits to an axis. A decision is made about polarity: head versus tail. Once that decision is stable, the rest of regeneration can unfold as an elaboration. But if that early decision is wrong, everything downstream becomes strange: two heads, no head, mis-scaled organs, or a body plan that seems to have forgotten what species it belongs to. The worm's later appearance depends on a very early computation about identity and direction.

From a human perspective, the temptation is to anthropomorphize. We want to say the fragment "knows" it needs a head. Taken literally, that is nonsense: knowledge implies a knower, and a tail fragment has no brain. But taken seriously as a claim about control and information, it is not nonsense at all. The fragment behaves as if it possesses a representation of a desired outcome and a way of reducing the difference between current and desired states. In other words, it behaves as a goal-directed system.

That goal-directedness is what makes the planarian more than a curiosity. It becomes a model for a broader thesis: intelligence is not confined to neurons, and problem-solving is not confined to conscious minds. A planarian's tissues coordinate over distances, decide among alternatives, and robustly achieve a target morphology. If such coordination is possible in a worm, then the roots of intelligence may lie deeper than brains—in the ways cells communicate, remember, and enforce collective goals.

There is a final twist to the blueprint idea. If the blueprint were merely passive—a plan stored somewhere—then regeneration would be the act of consulting it. But a planarian fragment does not consult a plan; it *stabilizes* a pattern. The blueprint is more like a memory held in a living circuit: an ongoing physiological state maintained by interactions among cells. It can be robust, but it can also be altered. A memory stored in a circuit is something you can, in principle, rewrite.

That possibility matters because it turns the planarian from a marvel into an experimental handle. If the blueprint is an active, physiological pattern rather than a fixed genetic script, then it might be possible to perturb it without changing the DNA. The next question, then, is not merely *how* fragments rebuild, but whether the instructions for rebuilding can be hacked: can we change the worm's target, persuading a tail to decide that it ought to grow another tail, or convincing a trunk that two heads are the new normal? The planarian's immortality, in other words, may depend on a kind of software—and software invites rewriting.

5.2 Rewriting the Pattern

Once you accept the planarian's most disorienting claim—that a scrap of tail can reliably decide to become a whole animal—the next question is almost irresistible: where is that decision written?

The obvious answer, taught to all of us in one form or another, is that it is written in genes. A cell has DNA; DNA encodes proteins; proteins build bodies. If you want a different body, change the genome. That story is not wrong, but in planarians it is incomplete in a very practical way. If genes alone fixed the body plan, then a trunk fragment and a tail fragment (genetically identical) would have no business making opposite choices about what to grow at a wound. Something other than DNA must be telling the same genome which chapter to read.

That "something" turns out to be both mundane and quietly radical: electricity. Not the crackling, dramatic electricity of lightning or electric eels, but the faint, continuous electrical differences that exist across every living tissue. Every cell is a tiny battery. Its membrane—the thin lipid boundary separating inside from outside—maintains a voltage. This *membrane potential* (often abbreviated as V_{mem}) is created by ion channels and pumps that move charged particles (ions) such as sodium, potassium, chloride, and protons across the membrane. We are used to the idea that nerve cells use voltage spikes to transmit signals. What is less familiar is that most cells maintain voltages too, and that slow, steady voltage patterns across tissues can act like a kind of body-wide information layer.

The phrase *bioelectric circuit* can sound like a gimmick until you picture what is physically present. Cells touch. Between them are tiny conduits called *gap junctions*: protein channels that allow ions and small molecules to pass directly from one cell to its neighbour. Link many cells through gap junctions and you create a conductive network. Add ion pumps and channels that set local voltages and you have, in effect, a living electrical circuit—not built of copper and silicon, but of membranes and proteins. The currents are minuscule. The patterns are slow. Yet they can be stable, and, crucially, they can store information in the way any network can store information: as a distributed state.

This is where the planarian becomes more than a regeneration marvel. It becomes a system you can *edit*.

A good way to feel the force of this claim is to imagine a rather different technology: a programmable thermostat. The house has hardware: walls, a heater, vents. But the comfort you experience depends on software: the target temperature. You can leave the hardware untouched and still change the outcome by altering the set point. If planarian anatomy has something like a set point—a target morphology—then the most powerful interventions would not be those that micromanage growth, but those that change what the tissue network thinks it is trying to achieve.

The experiments that opened this door are striking because of how gentle they can be. Instead of rewriting genes, researchers temporarily disrupt the electrical conversation among cells. In planarians, gap junction communication can be reduced for a short time using certain small molecules, as shown by Durant and colleagues in 2017. The animal is cut, exposed briefly to a gap junction blocker, and then returned to plain water. No transgenes. No permanent implants. No continued drug treatment. And yet, the worms can regenerate with the wrong body plan.

The most famous wrong body plan is as theatrical as it sounds: a two-headed planarian. A trunk fragment, which would ordinarily regenerate a head at one end and a tail at the other, instead produces heads at both ends. One might think this is merely a transient malformation, like a developmental accident. But the truly unsettling discovery is that the change can persist as a kind of *anatomical memory*, as Durant and colleagues reported in 2017. Some worms that regenerate with a normal-looking head and tail after treatment can still carry a hidden alteration. They look ordinary. They behave, for all casual purposes, like ordinary planarians. Yet when you cut them again later, in plain water, they are more likely than untreated controls to regenerate two heads. The first regeneration hides the edit; the next reveals it.

That is not how we are used to thinking about bodies. If you take a laptop, flip a setting deep in the firmware, and reboot it, you may not notice anything until you run a particular program—but the state was changed nonetheless. A planarian can be like that. The outward anatomy can appear normal while the underlying patterning system has been shifted into a different stable configuration. The term sometimes used for this is a *cryptic* pattern state: an altered internal instruction that is not visible until the system is challenged.

The implication is hard to escape. The body plan is not merely the product of gene expression in the moment. It is a stable, rewriteable physiological state distributed across tissues. The worm is not simply building; it is *remembering* what to build, and that memory can be edited.

A historical detour makes this even more interesting. Long before anyone spoke comfortably about bioelectric maps, scientists already knew that electricity could influence development. In the nineteenth century, embryologists and physiologists observed that injured tissues generated electrical currents, and that these could correlate with growth and healing. In the mid-twentieth century, as electrophysiology matured, the nervous system stole the spotlight because its electrical signals were dramatic and measurable. Developmental biology, meanwhile, became increasingly molecular. Genes, proteins, and chemical gradients offered a language of precise mechanisms, and the electrical aspect of non-neural tissues was often treated as background—a consequence rather than a cause.

Planarians helped reverse that neglect because they offered a behavioural readout in anatomy itself. If you disrupt gap junctions briefly and change whether a head appears where a tail should be, it is difficult to insist that the electrical network was merely a bystander.

The next question is whether voltage patterns are only permissive—allowing regeneration to proceed—or instructive, actually telling the tissue what to become. Here the evidence becomes more specific, and even more suggestive of a "software layer." Certain ion pumps and channels can be perturbed to bias the head-versus-tail decision. In planarians, inhibiting a particular ion pump (an H^+, K^+-ATPase, a molecular machine that moves protons and potassium ions across membranes) can derail head regeneration, as Beane and Levin showed in 2011. The details matter less than the logic: change ion transport, change voltage; change voltage, change what the wound becomes. In some experimental contexts, shifting the membrane voltage toward depolarization (making the inside of cells less negative relative to the outside) can favour head-like outcomes, while other manipulations can favour tail-like outcomes or impair proper patterning.

What makes this kind of intervention conceptually powerful is that it acts early. It does not wait for a head bud to appear and then distort it. It changes the initial decision-making landscape, the invisible prepattern that later growth follows. It is akin to changing the signposts in a city rather than pushing individual cars around.

At this stage, it is reasonable to worry that words like “map” and “prepattern” are just metaphors. Where is this map? What exactly is being mapped?

One answer, experimentally supported, is that planarians exhibit body-scale gradients of membrane potential. Not every region of the worm has the same electrical state. The head region, the trunk, and the tail can differ in characteristic ways. These differences are not simply local quirks; they form coherent, large-scale patterns that can extend over the entire body. If you could see voltage the way you see temperature in a thermal camera, you would not see a uniform worm. You would see an organism with electrical poles and regions, a physiological geography laid over its anatomy.

A second answer concerns connectivity. Gap junctions do not merely let ions leak; they couple cells into a network that can support stable global states. Anyone who has watched a crowd “choose” a chant or a stadium “choose” a wave has seen something similar at a different scale: local interactions can generate a coherent global pattern without a central controller. In tissues, a network of electrically coupled cells can settle into one of several stable configurations. Those configurations can act as memory states. Once the network is in one state, it tends to stay there, and when perturbed it tends to return. That is what makes it a plausible substrate for a target morphology.

Planarians offer a third, almost playful piece of evidence for discreteness. When the physiological network is perturbed in certain ways, the resulting heads are not merely "wrong" in a smooth continuum. They can fall into distinct, recognizable shapes, sometimes resembling head morphologies of other planarian species, as reported by Beane and colleagues in 2015. That is a remarkable claim, and it should not be misunderstood. It does not mean the worm has swapped genomes or literally become another species. It means that, from the same genetic starting point, the system can be nudged into producing alternative anatomical configurations that resemble the stable forms seen in related species.

This is exactly what you would expect if the body plan were governed by a set of attractors—stable outcomes available to the system—rather than by a single, rigid script. A piano can produce many notes from the same hardware; the note depends on which key you press. If the planarian has multiple "keys" in its physiological control system, then transient perturbations could press the system into a different key, producing a different but still coherent anatomical outcome.

The comparison to software does not mean the genome is irrelevant. Software runs on hardware. The available attractors, the stability of the network, the kinds of ion channels present, the strength of coupling between cells—all of this is genetically constrained. You cannot program a toaster to fly. But within the space allowed by the organism's evolved hardware, the physiological state can select among outcomes, maintain them, and sometimes hide them until the right perturbation reveals what has changed.

This view also helps resolve a tension with more familiar molecular explanations. Planarian regeneration is known to depend on gene regulatory pathways that establish anterior-posterior polarity and positional information. Chemical signals and gene expression patterns matter profoundly. The bioelectric story does not deny that; it rearranges the hierarchy. Rather than thinking of genes as issuing an unambiguous architectural command and physiology obediently executing it, we begin to see physiology as part of the control layer that sets the context in which gene networks operate. The same genes can be read differently depending on the global electrical state, because the electrical state shapes cell behaviour, coupling, and responsiveness.

A concrete analogy can keep this from becoming mystical. Consider a group chat. The participants (genes, proteins, cells) are the same people. But the tone and direction of the conversation can change depending on the channel settings: who is connected to whom, whether messages propagate quickly, whether certain members are muted, whether the chat is in "broadcast" mode or "free discussion." A small change to connectivity can radically alter the group's collective behaviour without changing any participant's identity. Gap junctions and membrane potentials are, among other things, ways of setting the tissue's communication mode.

Once you take that seriously, the two-headed planarian stops being a circus trick and becomes a diagnostic tool. It tells you that there exist global variables—distributed physiological states—that can override local wound cues. It tells you that tissues are not passive recipients of genetic instruction but active participants in a computation about form. And it tells you, perhaps most provocatively, that a body plan can be *rewritten* without being rebuilt from scratch.

The emotional resonance of this point is easy to miss if we keep it confined to worms. In human medicine, we often struggle with the opposite problem: tissues that grow when they should not, heal poorly when they should, or fail to restore complex structure after injury. If anatomy is controlled, even in part, by information stored in physiological networks, then the dream of regenerative medicine shifts. It becomes less about supplying cells and more about providing correct instructions. Instead of asking only, "Can we add stem cells?" we must also ask, "Can we restore the right pattern memory?" In the planarian, nature has already solved this in a way that is both robust and editable. That combination is the reason these animals keep returning to the centre of the story.

There is, however, a subtlety that prevents the bioelectric idea from collapsing into a simplistic slogan. If bioelectric circuits can be hacked, why does a planarian not constantly drift into bizarre forms? Why does it not routinely regenerate three heads or rearrange itself into a spiral? The answer seems to be that, most of the time, the patterning system is strongly stabilizing. It behaves less like a free-form artist and more like a conservative engineer. It corrects deviations. It pulls the system back toward a species-typical goal. The hacks work because they exploit moments of vulnerability—the early period after injury when the system is choosing among possible stable states.

That observation, in turn, points forward. If the organism has stable states, then regeneration becomes a problem in dynamics: how the system moves through state space, how it gets captured by one attractor rather than another, how it detects error and initiates correction. The language of maps and blueprints is helpful, but perhaps not deep enough. A better metaphor may be something like a valley in a landscape: roll the ball out of place and it rolls back; push it hard enough at the right time and it falls into a different valley, and from then on it returns there.

The planarian's most unsettling gift is not that it can regenerate. It is that it can be persuaded, briefly and gently, to adopt a different sense of what counts as "whole." A two-headed worm is not merely damaged; it has a different target.

That brings us to the next step. If anatomy can be stored as a stable, rewriteable state in a tissue-wide control network, then the right question is no longer "Where are the instructions?" but "What kind of memory is this?" Bodies, like minds, may have attractors, habits, and error-correcting loops. The planarian offers a chance to treat shape itself as a remembered goal—a form of *geometric memory* that tissues actively defend.

5.3 The Geometric Memory

A two-headed planarian is not only a creature with an odd face. It is a clue that the animal's body plan behaves less like a one-time construction project and more like a maintained commitment. Something in the tissue network can be flipped, and the worm will then keep rebuilding itself toward the new setting. That persistence forces a change in how we talk about shape. The familiar language of "development" suggests a historical event: an embryo builds a body, and the adult mostly preserves it. Planarians insist on something more active. Their shape is not merely an outcome. It is a goal.

The closest everyday analogy is not a blueprint pinned to a workshop wall, but a thermostat on the hallway. A thermostat does not create heat; it maintains temperature. When the room cools, it triggers a correction. When the room overheats, it shuts the heater down. The defining feature is not the machinery, but the *set point*—a remembered value the system works to preserve. When we say planarian anatomy behaves like a homeostatic goal, we are claiming that there exists, within the distributed physiology of the worm, something like a set point for geometry: a target morphology.

This sounds like poetry until you ask what regeneration actually requires. A fragment must not only produce missing structures but produce them in proportion. A head must end up the right size for the body it will control. A feeding organ must be placed where it can connect to the gut and the outside world. The body must regain bilateral symmetry. Most importantly, the process must stop. If regeneration were simply a growth programme switched on at a wound, it would be easy to imagine it overshooting—a worm that grows an enormous head, or an extra pharynx, or a tail that keeps extending. Instead, planarians typically return to a species-typical form with remarkable restraint, as if the system can sense when it is close enough to "correct" and then apply the brakes.

Homeostasis is an old idea in biology, and it has always carried a hint of intelligence. Claude Bernard, writing in the nineteenth century, argued that the stability of an organism's internal environment is the condition for free and independent life. Walter Cannon later named and popularized the concept of homeostasis, emphasizing the coordinated mechanisms that keep physiology within viable bounds. Most of that classic discussion concerns variables like temperature, pH, and glucose. Planarians broaden the discussion. They suggest that the organism's internal environment includes not only chemistry but *geometry*.

The difficult part is that temperature is a single number, easy to measure and easy to imagine a sensor tracking. Geometry is sprawling. A planarian's shape includes axes, proportions, organ placement, and tissue identities across space. What would it mean to "sense" a deviation in shape? And what would be doing the sensing, if not a brain?

The answer cannot be a single sensor. It must be distributed. Each region of tissue must contribute local information, and that local information must be integrated into a global assessment: how far are we from the target? This is precisely the kind of computation that networks are good at. A network does not store a single value in one place; it stores a state across many connections. When we speak of *geometric memory*, we mean that the worm's tissues collectively hold a memory of the correct body plan in their ongoing physiological configuration, and that this memory guides repair in the way a set point guides temperature regulation.

There is a concrete way to see that something like this is happening: planarians remodel, not just regrow. After amputation, the blastema forms at the wound, but the old tissue is not simply left alone as a passive scaffold. Existing structures change size and position relative to the new whole. The body re-scales. This is one of the quiet miracles of the system. A worm cut in half does not merely add a missing half; it becomes, over time, a smaller worm in which old tissues have been adjusted to fit the new scale.

This matters because it reveals that the target is not "rebuild what was removed" but "return to the correct overall configuration." Those are not equivalent goals. If you take a photograph, tear off the left half, and then glue a new left half to the remaining right half without resizing anything, you get a mismatched image: the seam is wrong, proportions are wrong, and the result looks like a collage. Planarians do not look like collages. They look like coherent organisms whose entire geometry has been recalculated.

That recalculation strongly suggests a feedback process. Feedback systems have a recognizable logic. They compare the current state to a desired state, compute an error, and act to reduce that error. The distinctive signature of feedback is robustness: the same goal can be reached from different starting points, and the system can correct for disturbances along the way. That is exactly what planarian regeneration looks like. A head fragment, a trunk fragment, and a tail fragment start from profoundly different conditions; they converge on the same species-typical form. When the system is perturbed, it often returns to the same end point, not because it is rigid, but because it is correcting.

The idea of an "end point" is where the language of dynamical systems becomes helpful. Imagine a landscape of hills and valleys. A marble placed on a slope will roll downward until it settles in a valley. The valley is an *attractor*: a stable state toward which the system tends. In this picture, each possible body plan corresponds to a valley in an abstract landscape of physiological states. The normal planarian morphology is a deep valley: most perturbations roll back into it. But there may be other valleys. A two-headed morphology may be another, shallower valley that the system can be pushed into under special conditions. Once there, the network may remain stable and, when injured, roll back into that same valley again. That is what it would mean for anatomy to be a memory state.

The experiments described earlier, where a transient disruption of gap junction communication can produce a stable tendency to regenerate two heads, make more sense in this framework. The temporary perturbation is not instructing every cell "make a head here." It is nudging the system into a different basin of attraction. After the perturbation ends, the system is no longer being forced; it is simply following the logic of its new stable state.

The cryptic cases—worms that look normal yet later regenerate abnormally—fit even more neatly. They suggest that an attractor can be hidden in plain sight. Two different internal network states can produce nearly identical outward anatomy under unchallenged conditions, much as two different computer configurations can both display the same desktop until you run a program that depends on the deeper settings. The anatomy is not the whole story. The memory lies in the control system that produces anatomy when needed.

This is also why "software" is not just a flashy metaphor. Software is not visible in the appearance of a device. A phone can look the same before and after an operating system update, yet behave differently when tested. Similarly, a planarian can look the same while carrying a different patterning regime. The organism's geometry is, in part, the expression of an internal state that can remain latent.

All of this might still feel uncomfortably abstract, so it helps to anchor it in the kind of evidence that biologists take seriously: specific interventions with specific outcomes. Manipulating ion pumps and channels can bias whether an anterior wound becomes a head. Temporarily blocking gap junctions can shift the polarity decision and produce stable changes in regenerative outcomes. Altering physiological connectivity can lead to discrete alternative head morphologies. In each case, the common thread is not simply that electricity matters, but that the tissue network supports *multistability*: more than one stable pattern state is possible, and the system can be switched among them.

Multistability is familiar in other biological contexts. The immune system can be tolerant or inflammatory; metabolic networks can settle into different regimes; neurons can be excitable or quiescent. In each case, the system's behaviour depends not only on components but on the state of the network. Planarians invite us to add a new member to that list: body plan itself.

This is where we can return, briefly, to the more human-sounding word *memory*. The planarian's geometric memory is not a recollection of events. It is a retained target. It is closer to a habit than a story: a tendency to settle into a particular form, and to correct deviations from it.

The temptation now is to ask whether planarians have memory in the ordinary psychological sense as well. Here we enter a chapter of scientific history that is both fascinating and cautionary. In the mid-twentieth century, James McConnell became famous for claims that planarians could be classically conditioned and that the memory could persist through regeneration, even through cannibalism—a trained worm eaten by an untrained worm, transferring memory as if it were a nutrient. These ideas captured the public imagination, in part because they flirted with a deliciously scandalous possibility: memory as a transferable substance, perhaps even a form of "memory RNA." They also generated intense controversy, because many laboratories had trouble reproducing the effects, and the experimental methods were often criticised as subjective or insufficiently controlled.

The story matters for our purposes not because the cannibalism claims should be revived as fact, but because it highlights how easy it is to slide from a genuine mystery to a seductive narrative. Planarians are the kind of organism that tempts grand claims. If you want to argue that memory is stored everywhere, that identity is fluid, that information is chemical and transferable, planarians will cheerfully play along. They are a mirror that reflects our intellectual desires.

Modern behavioural work has tried to bring sobriety to the question by using automated training and testing paradigms that reduce human bias. Some experiments report that trained responses can show a "savings" effect after decapitation and head regeneration: the worm relearns faster than a naive worm, suggesting that something of the prior training persists through the loss and regrowth of the head. Even if one accepts these results, they do not imply that memory is a mystical fluid. They raise a more grounded possibility: information relevant to behaviour might be stored not only in the brain but also in peripheral tissues, or in stable physiological states that can influence the rebuilding of neural circuits. The planarian, with its ability to regrow a brain, becomes a natural test case for the broader theme of this book: intelligence and memory may be properties of networks, not of organs.

Still, it is important to keep the two kinds of memory distinct. Behavioural memory is about learned associations. Geometric memory is about anatomical goals. They may be related in deep ways, but they are not the same phenomenon. The safest and most fruitful claim, supported by the clearest evidence, is that planarians exhibit a robust geometric memory: a persistent, rewriteable target morphology that guides regeneration through distributed control.

Once you see shape as a homeostatic goal, several puzzles rearrange themselves. One puzzle is why regeneration is so reliable. In a feed-forward system, reliability requires precise execution of a sequence; any early error can cascade. In a feedback system, reliability can be achieved by continual correction. Planarians look like the latter. Another puzzle is why tissues can remodel long after the initial injury. In a feed-forward script, the programme ends; in a feedback regime, the system keeps checking. A third puzzle is why subtle physiological perturbations can have dramatic effects. In a feedback system with multiple attractors, small nudges at the right time can push the system into a different basin, after which the system's own corrective behaviour will stabilize the new outcome.

None of this requires a brain. It requires a control architecture: sensors (local detection of state), communication (signals passing across distances), and effectors (cell behaviours that change tissue). In planarians, those ingredients exist in the ordinary machinery of cells. Ion channels set voltages. Gap junctions couple regions. Gene networks and cellular mechanics execute the changes. The surprising part is not that such components exist, but that they can be organized into a system that treats anatomy as a variable to regulate.

This perspective also changes how we think about errors. In most animals, severe injury is an error the body cannot correct; the best it can do is patch. Planarians correct. The "error" is not merely damage; it is deviation from a geometric target. The body acts as if it can measure that deviation and reduce it over time. The organism, in other words, exhibits agency at the level of form: it controls outcomes, steering matter toward a remembered configuration.

If you accept that, then the two-headed worm becomes less a monster and more a proof of principle. It shows that the target can be changed. It suggests that the space of possible forms may be broader than we assume, and that evolution may have shaped not just the parts of organisms but the attractor landscapes that their tissues inhabit.

A final thought presses in, uncomfortable and exhilarating. If planarian tissues can store a geometric set point and enforce it through distributed feedback, then shape is not a static inscription of genes. It is a dynamic agreement among cells about what the organism is supposed to be. Agreements can be stable, but they can also be negotiated, corrupted, repaired, or rewritten. The planarian is a creature that lives by such agreements.

That takes us beyond worms. A human body also maintains a target morphology in countless small ways: wounds close, bones remodel, organs maintain size, and tissues resist change. We call these processes healing and homeostasis, but we rarely think of them as memory. Planarians force that conceptual shift. They make it plausible that the control of form is a kind of cognition performed by tissues.

The next step is to ask how far this idea travels. If a worm can remember its shape, and if that memory can be edited by changing the electrical conversations among cells, then perhaps the boundary between *development* and *computation* is thinner than we thought. The worm that never forgets its geometry may be only the opening example of a more general principle: life solves problems by storing goals in the very networks that build bodies.

Chapter 6

When the Collective Breaks

Multicellular life relies on a fragile negotiation, requiring billions of cells to subordinate their selfish drives to a larger anatomical purpose. When the bioelectric conversation sustaining this cooperation fails, cells do not merely break; they revert to an ancient, unicellular survival mode, shrinking their cognitive horizons to focus on self-replication rather than the whole.

6.1 The Defector's Dilemma

A human body is not a single thing so much as a long-running agreement. Trillions of cells, each with the full biochemical machinery to eat, grow, divide, move, and defend itself, spend a lifetime doing none of those things *as individuals*. Instead, they accept roles: a liver cell detoxifies, a neuron signals, a skin cell forms a barrier, an immune cell patrols. They also accept limits: divide only when asked, die when obsolete, share resources, keep the neighborhood clean. Multicellular life is, in that sense, an act of sustained cooperation.

Calling it a "social contract" may sound like an overreach—cells do not negotiate, feel loyalty, or read constitutions. Yet the analogy earns its keep because it captures something that ordinary mechanical metaphors miss: the body has to solve a permanent political problem. Every cell carries an old temptation. For most of Earth's history, the winning strategy for a cell was simple: replicate as quickly as conditions allow. Multicellularity did not erase that ancient imperative; it domesticated it. The organism survives only if individual cells agree, continually, to subordinate their local advantage to a collective goal.

Cancer, in this framing, is not merely "bad luck in the genome" or a list of mutations to catalogue. It is what happens when the agreement fails. A cell (or a small group) stops treating the organ's priorities as binding and starts acting like an autonomous organism again. It defects.

To see why defection is such a deep hazard, it helps to notice that cooperation is not the default state of matter. The body must *actively* maintain it, moment by moment, using layers of communication and enforcement. Growth factors tell cells when to divide. Contact inhibition tells them when to stop. Immune surveillance checks identity papers. Tissue architecture imposes physical constraints. And, threaded through much of the body, there is a quieter and older channel of conversation: direct electrical and chemical coupling through gap junctions.

Gap junctions are microscopic bridges between adjacent cells. Each bridge is built from protein subunits (connexins, in vertebrates) that align to form a pore connecting the cytoplasm of one cell to its neighbor. Through these pores pass ions and small molecules: the kinds of signals that say, without words, "we are part of the same tissue; we share a fate." This is not the crackling electrical communication of neurons, but a distributed, steady exchange. If neurons are a body's fast email, gap junctions are more like shared spreadsheets—a way to keep local accounts synchronized.

When that synchronization is robust, cells behave less like independent contractors and more like a disciplined workforce. When it is lost, cells become informationally isolated. They still receive some messages from their surroundings, but they no longer participate in the fine-grained, peer-to-peer consensus that helps hold a tissue in a coherent state. In that informational loneliness, defection becomes easier.

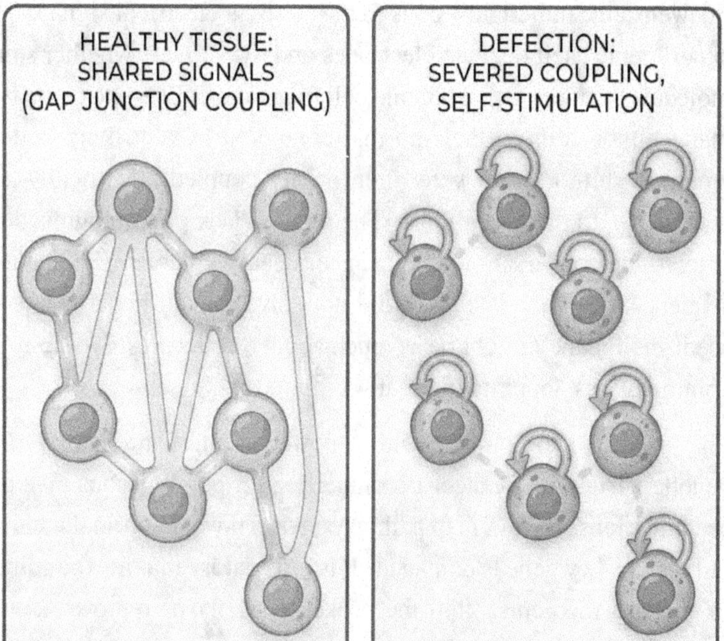

The point of the diagram is not to claim that cancer is *simply* "gap junctions broken." Many cancers show altered coupling, but the relationship is complex: connexins can behave like suppressors in one context and accomplices in another. The diagram's purpose is conceptual. It asks you to picture the tissue not as a heap of parts but as a communicating network. In a network, failure is not only about damaged nodes; it is also about broken edges. The body's intelligence depends on those edges.

A historical anecdote makes this intuition concrete. In the 1960s, long before the modern molecular era had turned cancer into a story of oncogenes and tumor suppressors, a physiologist named Werner Loewenstein studied how cells in tissues pass electrical signals. His experiments used delicate electrodes and dyes to test whether small molecules could move from one cell to another. When he compared many tumor cells to their normal counterparts, a striking pattern emerged: tumor cells were often poorly coupled, electrically and chemically, to their neighbors. The phrase "lack of communication" entered the conversation about cancer decades before anyone could sequence a tumor. The work did not solve cancer, but it planted a seed: malignancy might be as much about a cell's *relationship* to its community as about its internal wiring.

That seed had to compete with a powerful alternative story. The genetic narrative of cancer is compelling for good reasons. We can see mutations; we can map pathways; we can watch normal controls fail when a key gene is damaged. It is also a story that fits the culture of modern medicine: find the broken part, fix or remove it. But the social-contract view insists on a different emphasis. A cell with dangerous mutations is like a citizen with dangerous tools. Whether disaster follows depends, in part, on whether the surrounding society can still constrain behavior.

This is not hand-waving. Even in everyday life, you already accept that context can dominate intrinsic potential. A person with a violent impulse is not the same threat in a stable community as in a collapsing one. A teenager in a supportive school is not the same risk as in an environment that rewards recklessness. The same gene can mean different things in different bodies; the same mutation can have different consequences depending on tissue organization, immune state, inflammation, and mechanical stress. We are not replacing genetics with sociology; we are admitting that the organism is a *multilevel* system, where causation runs through relationships.

The problem is old enough to have shaped evolution. Multicellularity is one of life's "major transitions"—a jump in complexity that required mechanisms to suppress cheating. If you imagine the earliest multicellular collectives as loose gatherings of cells, the easiest way for such a collective to fail is obvious: one cell could take the benefits of the group (protection, shared resources) while ignoring its costs (restraint, specialization). Over time, successful multicellular lineages evolved a battery of constraints that make defection hard: regulated proliferation, programmed cell death (apoptosis), differentiation (division of labor), resource allocation rules, and maintenance of the extracellular environment. Cancer looks, in many respects, like a systematic undoing of these constraints. Cells ignore stop signals. They evade cell death. They de-differentiate, abandoning their specialized job for a more primitive, flexible state. They hoard nutrients and secrete enzymes that remodel their surroundings.

This is where the "defector's dilemma" becomes more than a metaphor. In game theory, cooperation can be stable if defectors are punished, if cooperators preferentially interact with each other, or if the benefits of cooperation depend on reputation and repeated interactions. Tissues implement versions of all three. Neighboring cells constantly monitor each other's behavior through chemical cues and physical contacts. Abnormal cells can be eliminated by immune cells or by a process sometimes called cell competition, in which healthier cells push out their less fit neighbors. And many tissues keep a reserve of stem cells that can replace cells that must be removed. These are not moral systems. They are enforcement systems.

Cancer emerges when enforcement fails, and that failure can happen in multiple ways. A cell may become harder to punish by disabling apoptosis. It may become harder to recognize by changing surface markers. Or it may exploit the tissue's own enforcement mechanisms—for instance, by creating chronic inflammation that keeps the local environment in a wound-like state, rich in growth factors meant for repair. A wound is cooperation under emergency rules; cancer is often cooperation's emergency rules stuck on.

The social-contract frame also changes how we think about "default" behaviors. A famous debate in cancer biology contrasts two intuitions. One says that normal cells are naturally quiet and that cancer arises when genetic damage presses the accelerator. The other says that cells, like all living things, have an inherent drive to proliferate, and that normal tissue is what holds that drive in check through organization and signals. In the latter view, the central miracle is not that cancer cells grow; it is that most of your cells *do not.* They are constantly being talked out of a life strategy that would have been perfectly sensible for their distant unicellular ancestors.

This is where gap junctions return as more than a detail. A tissue is not only a pile of cells receiving hormones from the bloodstream. It is also a local democracy of electrical and chemical states. When cells share ions, they help set a common "resting potential" landscape—a pattern of voltages across a tissue that influences whether cells divide, migrate, or differentiate. These voltage patterns are not mere by-products. In many organisms, they are instructive signals during development, guiding large-scale anatomy without a nervous system. A developing embryo uses bioelectric cues the way a city uses zoning laws: not to dictate every molecule, but to constrain what kinds of structures can form where.

If a tissue's bioelectric conversation helps enforce the multicellular contract, then severing that conversation shrinks the reach of the tissue's governance. The cell becomes, informationally, more alone. Alone cells are more likely to revert to older priorities: survival now, replication now, local advantage now. You can begin to see why cancer so often looks like a loss of "memory" of the body plan. It is not only that a gene breaks. It is that the cell's world becomes smaller.

A concrete example helps anchor this. Consider the classic observation that many cancers resemble poorly differentiated tissue—pathologists describe tumors as "undifferentiated" or "anaplastic" when they have lost the distinctive architecture of their tissue of origin. A normal gland looks like an organized array of structures; a malignant glandular tumor can look like a chaotic mass. That chaos is not purely visual. Differentiation is a cell's commitment to a job, and a job is a promise to the collective. When cells abandon specialization, they stop producing the goods the tissue depends on and resume behaviors useful to a free-living cell: fast division, stress resistance, migration. In social terms, they stop being bakers or builders and become opportunists.

Why does de-differentiation travel with malignancy so often? Part of the answer is genetic: mutations can disrupt differentiation pathways. But part of it is informational: differentiation is maintained by continuous cues from the tissue. Remove the cues, and the cell's state can drift. If you have ever tried to keep a habit without supportive surroundings, you understand the principle. In biology, the "supportive surroundings" include chemical gradients, mechanical stiffness, and electrical coupling. A differentiated cell, in many tissues, is not a permanently altered object; it is a pattern sustained by feedback.

Here is the twist that makes the social-contract view emotionally unsettling. If cancer is defection, then the enemy is not wholly alien.

It is not a parasite invading from outside. It is a citizen using the body's own resources against it, exploiting loopholes in the rules, and recruiting the neighborhood to support a new regime. Tumors are notorious for persuading blood vessels to grow toward them (angiogenesis), for altering immune responses, for reshaping surrounding connective tissue. A successful defector does not merely ignore society; it often rewrites society.

This makes prevention and therapy feel less like fixing a broken gadget and more like maintaining a fragile civic order. It also explains why the same carcinogen can have different consequences in different tissues, and why the same mutation can be present without cancer. The organism is not passive. It is an active, distributed intelligence that normally keeps cellular self-interest aligned with the body's long-term goals.

And yet, despite all this enforcement, cancer is common. That is not a failure of design; it is the cost of being made of living agents. The very features that make multicellular life flexible and repairable—cells that can divide, migrate, change identity, respond to stress—also provide avenues for defection. A body made of inert bricks would not get cancer. It also could not heal.

The defector's dilemma, then, is the permanent tension at the heart of complex life: you want cells capable of initiative, because initiative underlies development, regeneration, and immune defense. But initiative creates the possibility of rebellion. Evolution did not solve this dilemma once and for all; it built overlapping compromises.

This framing prepares us for a subtler question: what exactly is lost when a cell defects? We can list the molecular pathways and the broken checkpoints, but those are the *mechanisms*. The deeper loss is informational. A healthy tissue acts as if it can "see" beyond the immediate moment—as if each cell's decisions incorporate distant goals, like maintaining an organ's shape or preserving a boundary. A tumor behaves as if its future has collapsed into the present. Its decisions look locally rational and globally disastrous.

That collapse of perspective is the next step in the story. Once we have cancer in mind as a breakdown of cooperation, we can ask how cooperation enlarges a cell's horizon in the first place, and how defection shrinks it. The next section follows that shrinking boundary—the way a tumor seems to pull its world inward, trading the body's long-term plan for an aggressively short-sighted kind of competence.

6.2 Shrinking the Event Horizon

When a cell defects from the body's social contract, it does not simply become "more selfish" in the moral sense. Something more precise happens: its sense of *relevance* collapses. Signals that once mattered stop mattering. Distant consequences cease to constrain present choices. The cell's world shrinks until only the nearby and the immediate feel real.

Physics has a phrase for a boundary beyond which influence cannot reach: an event horizon. Outside it, events exist, but they cannot affect you. Inside it, everything counts. Tissues also have horizons, not set by the speed of light but by the reach of communication and feedback. A healthy organ is full of long-range constraints: growth is not merely local appetite; it is tied to organ size, architecture, repair needs, and the body's overall state. Cells "know" about these distant goals not because they are conscious, but because they are embedded in networks that make far-away conditions locally legible. Hormones carry news from endocrine glands. Nerves tune blood flow. Immune cells report damage. Neighbors share ions and small molecules. The extracellular matrix—the supportive meshwork between cells—acts like both scaffolding and signal board. From these many channels, a cell continuously infers: *What kind of place am I in? What is the organ trying to maintain? What is permitted here?*

Cancer can be read as a narrowing of that inference. The tumor cell's "cognitive light cone"—the slice of past and future, near and far, that effectively shapes its decisions—contracts. The cell becomes brilliant at surviving a hostile, hypoxic (low-oxygen), acidic pocket of tissue. It becomes indifferent to what that pocket does to the organ, the organism, or the future. Many of cancer's most famous traits make sense as strategies for life inside a shrunken horizon.

Begin with the simplest example: uncontrolled proliferation. "Uncontrolled" can mislead, as if cancer were chaos. Tumors are often organized around local logic. They expand where nutrients and space permit, and they slow where constraints tighten. Their growth resembles an invasive plant in a garden: not random, but indifferent to the gardener's plan. In a healthy tissue, division is a public act, licensed by communal signals. In a tumor, division becomes a private decision, justified by the immediate availability of resources and by internal circuits that amplify "go" signals and ignore "stop" signals. The cell is no longer solving the organ's problem; it is solving its own.

One can push the analogy further. In the previous section, the social contract depended on communication among cells. If you cut communication, you do not merely remove information; you remove *accountability*. In a well-coupled tissue, a cell's internal state is continuously compared against the states of its neighbors. Deviations become visible, and visibility makes correction possible. A cell that is too depolarized (with an abnormally reduced electrical voltage across its membrane), too stressed, too proliferative, too out of place, can be nudged back or eliminated. But if the coupling is broken—if junctions close, if signaling pathways are rewired, if the surrounding matrix becomes distorted—then the comparisons weaken. The cell's internal feedback loops become dominant. A self-reinforcing circuit can masquerade as truth.

This is a general principle in complex systems: the smaller the horizon, the more tempting positive feedback becomes. Positive feedback means that an action makes itself more likely in the future. A microphone too close to a speaker screeches because sound amplifies sound; the system can no longer listen to the room. Many cancers behave like a tissue that has moved its microphone too close. Autocrine signaling (cells secreting growth factors that act back on themselves), chronic activation of proliferation pathways, and local remodeling of the tumor microenvironment all create loops that amplify local advantage while muting global context.

There is a vivid historical precedent for thinking this way, long before "microenvironment" became a fashionable word. In the early twentieth century, the embryologist C. H. Waddington introduced the metaphor of an "epigenetic landscape" to explain how cells, during development, roll into stable fates. Picture a marble on a hilly surface: it can fall into different valleys, and once in a valley it tends to stay there. Waddington's point was not that cells are marbles, but that development is constrained by a landscape of stabilizing interactions. The cell's fate is not dictated by a single gene but by a network that funnels possibilities.

Cancer can be imagined as a landscape whose valleys have been reshaped. A differentiated cell sits in a deep valley: stable, specialized, hard to dislodge. Injury, inflammation, or stress can shallow that valley, making it easier to slip into more primitive states. Mutations can also flatten the terrain. But importantly, the landscape is not inside the cell alone. It is carved by the tissue. A stiffened matrix, altered oxygen levels, disrupted electrical coupling, and persistent inflammatory cues can all remodel the hills. Under those conditions, the same genome can yield a different stability. The "event horizon" shrinks because the broader landscape that normally channels behavior is no longer reliably sensed or enforced.

At this point, one might object: cancer clearly responds to its environment. Tumors recruit blood vessels, evade immune attacks, and adapt to therapies. How can that be called a shrunken horizon? The answer lies in the difference between *responsiveness* and *integration*. A thermostat is responsive. It senses temperature and turns a heater on or off. But it is not integrated into the goals of a household; it will happily warm an empty house all day. Likewise, a tumor can become exquisitely responsive to local cues and still be catastrophically unintegrated from the organism's goals. Shrinking the horizon does not mean losing sensitivity; it means losing the ability to treat distant constraints as binding.

In healthy development and regeneration, cells routinely behave as if they are optimizing for outcomes they cannot possibly "see" directly: the correct number of fingers, the closure of a wound without overgrowth, the maintenance of a smooth epithelial sheet. They achieve this by participating in collective fields of information. Those fields can be chemical (morphogen gradients), mechanical (tension and stiffness), or electrical (voltage patterns across tissues). The field encodes the body plan as a set of distributed cues that each cell reads locally.

Bioelectric fields deserve special attention because they are both ancient and easy to overlook. Every cell maintains a voltage difference between its inside and outside, created by ion pumps and channels in its membrane. Neurons use rapid changes in this voltage to communicate, but non-neural tissues use slower, steadier voltage patterns as a kind of spatial language. A region of tissue can be relatively depolarized or hyperpolarized compared to its neighbors, and this difference can influence whether cells proliferate, migrate, or differentiate. The crucial point is that voltage patterns can extend over distances of many cell diameters, providing a medium for long-range coordination without nerves.

A concrete example comes from work in frog embryos, where researchers introduced human oncogenes—genes that, when activated inappropriately, can drive cancer-like behavior—and observed tumor-like structures. The surprising twist was that the bioelectric state of cells *far from* the prospective tumor site could influence whether those structures formed. When distant tissues were experimentally hyperpolarized (made more negative inside relative to outside), tumor incidence dropped, even though the oncogene was still present. In other words, the "field" in which the oncogene operated mattered. The embryo's long-range electrical conversation could overrule, or at least strongly modulate, a powerful genetic push toward malignancy.

For our purposes, this is not merely a curiosity; it is a clue to what the event horizon really is. A cell's horizon is the radius over which the system's collective state can constrain its behavior. In a healthy embryo or tissue, that radius can be surprisingly large. In a tumor, it tends to collapse. The tumor becomes a pocket of tissue that is electrically, chemically, and mechanically out of sync with its surroundings, a sort of local jurisdiction with its own rules.

You can feel the consequences of this collapse when you look at what tumors do to time. Many cancers activate stress responses that are excellent in the short term but ruinous in the long term. They rely heavily on glycolysis (a way of generating energy from glucose) even when oxygen is available, a phenomenon historically associated with Otto Warburg in the 1920s. Warburg proposed that cancer is fundamentally a problem of altered metabolism, pointing to the tendency of tumor cells to ferment glucose into lactate. Modern biology has complicated Warburg's original causal claim, but his observation remains a cornerstone: tumors often adopt metabolic strategies that support rapid growth and survival in poor conditions, even at the cost of efficiency and long-term tissue health. It is as if the cell is living on credit, spending resources and generating waste in ways that make sense only if the future is discounted.

Warburg's era offers a useful historical anecdote because it reminds us that cancer has always tempted scientists to pick a single "root cause." For a time, metabolism seemed the answer; later, genes seemed the answer; then signaling pathways; then microenvironments; now the microbiome and immune context. The event-horizon view does not deny any of these. It tries to supply the missing grammar that unites them: each "cause" is a way of changing what the cell can effectively measure and what outcomes it can effectively control. Metabolism changes what resources are locally available and what stresses are endured. Mutations change what signals the cell can hear and how strongly it amplifies them. Microenvironments change the landscape that defines stable behaviors. Bioelectric fields change the long-range coherence of the tissue. Each can shrink the horizon, and once the horizon is small, many different molecular routes can converge on a similar malignant logic.

This helps explain another puzzling feature of cancer: its ability to converge. Different tumors carry different mutations, arise in different organs, and face different pressures, yet they repeatedly reinvent a familiar set of behaviors: proliferation, invasion, angiogenesis, immune evasion, altered metabolism. The classical way to interpret this is to list the "hallmarks of cancer." The event-horizon way is to ask what kinds of behaviors are favored when a cell's effective world becomes local. If you cannot rely on communal infrastructure, you build your own supply lines (angiogenesis). If you cannot trust external signals to keep you alive, you block apoptosis. If your neighborhood becomes hostile, you migrate. If immune cells threaten you,

you cloak yourself or manipulate them. The pattern looks less like a random accumulation of tricks and more like a coherent survival program for life outside the multicellular contract.

Some theorists have suggested that this coherence reflects an "atavism": a reversion toward older, unicellular-like programs that were never truly erased from the genome. The word can be misleading if taken too literally. Cancer cells are not time travelers, and they are not simple bacteria in disguise. But as a narrative scaffold, atavism highlights something real: many genes and pathways used in cancer are ancient stress-response and proliferation modules conserved across evolution. In a healthy multicellular context, those modules are fenced in, recruited for wound healing, regeneration, and immune defense. When the fences fail, the old modules can dominate. The cell's priorities begin to resemble those of a free-living organism.

Notice how well this fits the idea of a shrinking horizon. A unicellular organism has no obligation to distant anatomy. Its "body plan" is mostly the boundary of its membrane. Its time horizon is short: survive, divide, persist through stress. A multicellular organism extends horizons by building communication networks that tie local actions to global goals. If those networks fray, the local agent falls back to what its circuitry already knows how to do. Cancer, in this view, is not the invention of a new intelligence but the reassertion of an old one, operating under impoverished information.

At the emotional level, this reframes the tragedy of cancer. We often imagine cancer as pure error: a broken machine producing garbage. The event-horizon view paints something more haunting: a competent system solving the wrong problem. Tumor cells are not merely malfunctioning; they are adapting. They are learning their local environment, exploiting gradients, negotiating with immune cells, diversifying under stress. They are, in a narrow sense, intelligent.

The horror is that this intelligence is trapped in a tiny world, unable to "care" about the organism because the signals that would make the organism real have been cut, drowned out, or reinterpreted.

This also clarifies why some treatments backfire. Therapies that brutally attack a tumor can select for clones that are even better at local survival: more stress-resistant, more invasive, more able to hide. That is not a moral critique of therapy; it is an ecological fact. If you push a system whose horizon is already small, you may further reward strategies optimized for immediacy. In contrast, approaches that restore constraints—that enlarge the horizon, reconnect the cell to tissue-level information—might, in principle, shift what strategies are favored. Instead of trying to outgun a cell's local survival intelligence, one might try to give it back a world big enough to include the body plan.

That possibility can sound fanciful. Yet development itself is full of examples where context overrides dangerous potential. Embryos can correct surprising perturbations; tissues can normalize cells that would otherwise behave badly; architecture can dominate genotype. These are not miracles. They are the everyday power of collective control systems that operate above the level of single cells.

The immediate question, then, is practical: if cancer involves a shrunken event horizon, can we *expand* it again? Could a tumor cell, still bearing many mutations, be coaxed into rejoining the tissue's consensus? Could one "talk" cells back into the fold, not by persuading them in words, but by restoring the electrical, chemical, and mechanical conversations that make the organism's goals locally present?

The next section turns to experiments that attempt exactly that: interventions that do not merely poison tumor cells, but alter the informational environment—especially bioelectric signals—so that even damaged cells behave as if they remember the body plan.

6.3 Talking Them Back into the Fold

If cancer is only a matter of broken hardware—a cell's genes irreparably damaged—then the only sensible response is demolition. Cut it out, burn it, poison it, starve it. Much of oncology has been built on that logic, and for many cancers it saves lives. Yet the picture developed in the previous pages suggests a second possibility. A defector is not always a foreign invader; sometimes it is a citizen who has stopped receiving the messages that make citizenship meaningful. If that is even partly true, then killing may not be the only way to restore order. Some cells might be *reintegrated.*

That word can sound naïve when applied to a disease that still terrifies. But "reintegrate" does not mean forgiving a tumor or pretending mutations do not matter. It means asking whether a malignant phenotype—the observable behavior of a cell: uncontrolled growth, invasion, disregard for tissue architecture—is always a permanent consequence of its internal damage, or whether, in some contexts, it can be overwritten by stronger, tissue-level instructions. In everyday life, behavior can be remarkably context-dependent; the same person can be reckless in one setting and responsible in another. Biology, too, often treats phenotype as something negotiated between internal machinery and external constraints.

A historical anecdote makes this idea feel less like wishful thinking. In the 1970s, experiments with mouse teratocarcinomas (tumors arising from germ cells) produced a startling result. When malignant cells from these tumors were introduced into an early mouse embryo (a blastocyst), some of them contributed to normal tissues in the resulting animal, as Mintz and Illmensee showed in 1975. The cancerous cells did not remain a rampaging mass; in the right developmental environment, they behaved like ordinary building blocks. The genes did not magically revert to pristine form. What changed was the *conversation* those cells were forced to join: the embryo's powerful patterning signals and architectural constraints.

Not all cancers can be tamed this way, and no one is proposing that we treat human tumors by inserting them into embryos. The point is conceptual and durable: malignancy can be, at least sometimes, a state that the collective environment can suppress. The body plan can dominate genotype.

A parallel story emerged later from the world of tissue engineering and breast cancer research. Cells grown on a flat plastic dish behave differently from cells embedded in a three-dimensional scaffold that mimics real tissue. Malignant breast cells that look aggressive in two-dimensional culture can, under certain conditions in three-dimensional matrices, form organized, growth-arrested structures resembling normal glandular units, as Weaver and Bissell's group showed in 1997. The changes involve familiar molecular players—integrins (cell—matrix adhesion receptors), growth factor signaling, and downstream pathways such as MAP kinase. But the deeper lesson is again about context. Architecture is not decoration; it is instruction. A tissue is not merely a container for cells; it is a regulatory field that can stabilize or destabilize cellular identities.

These examples broaden the question from "How do we kill cancer cells?" to "How do we restore multicellular governance?" Once asked in that way, bioelectricity becomes hard to ignore. Every cell sits in an electrical landscape. Ion channels and pumps set a resting membrane potential, and tissues develop patterns of voltage that extend beyond individual cells. These patterns are not just the nervous system's special trick; they are a general medium for coordination in living matter. If malignancy involves a shrinking event horizon—cells trapped in local feedback—then a tissue-scale electrical signal is exactly the sort of intervention that could, in principle, expand that horizon again, reminding cells that they are part of something larger.

One set of experiments, especially vivid in frog embryos, captures the logic. Researchers can introduce human oncogenes into developing tadpoles and induce tumor-like structures: disorganized growths that recruit immune cells, acidify their surroundings, and display several hallmarks familiar from human cancers, as shown by Chernet and Levin in 2014. That alone is striking: the machinery of malignancy is conserved enough that a human gene can derail a frog's developing tissues. But the more surprising result comes when the experimenters do *not* attack the tumor directly. Instead, they manipulate the bioelectric state of cells at a distance by expressing specific ion channels (often via injected RNA). By shifting the membrane potentials of non-tumor cells—hyperpolarizing them so that their interiors become more negative relative to the outside—they can reduce the incidence of oncogene-driven tumor-like growths, and in some settings suppress already initiated abnormal structures. The tumor is being governed by a field, not only by its own genome.

This is the kind of result that can be misheard. It does not mean that "electricity cures cancer" in the simplistic sense of zapping tumors. It does not mean that bioelectricity replaces genetics or immunology. It means that tissues possess *set points*—preferred, actively maintained electrical states—and that these set points are part of the body's pattern-control system. When the set points are restored, cells that were poised to defect may find themselves back under collective constraint. The "software" of the tissue, distributed across many cells, can sometimes compensate for damaged "hardware" in a subset of them.

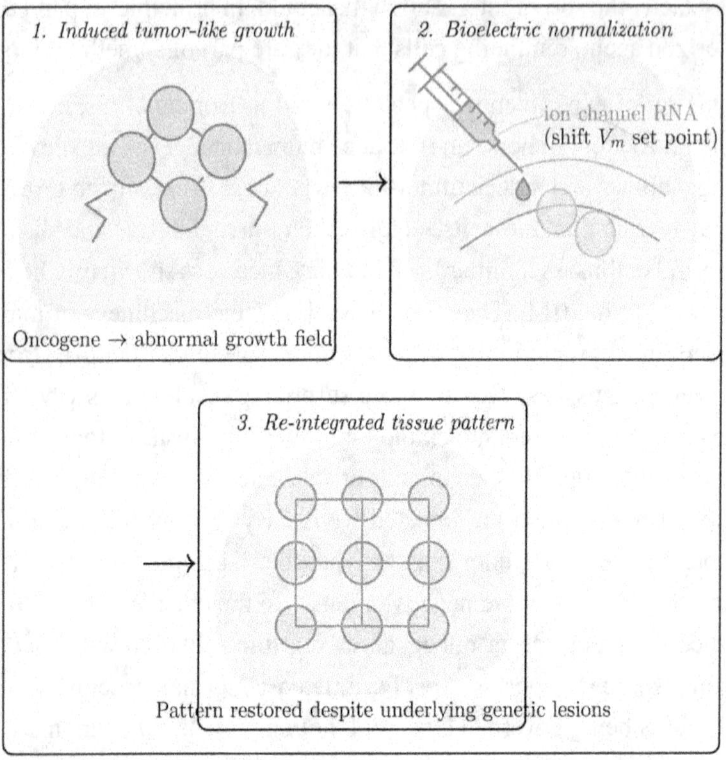

The diagram is a simplification, but it captures the conceptual structure. The first panel is a local breakdown: a tumor-like cluster establishes its own self-supporting conditions. The second panel is the intervention: not a toxin targeted at the cluster, but a change to a broader electrical environment that alters what "normal" means in that tissue. The third panel is the desired outcome: not the annihilation of every suspect cell, but the re-establishment of organized structure.

This is a different therapeutic imagination. Traditional cancer therapy often assumes that the tumor is a self-contained enemy, and the rest of the body is a battlefield. A field-based view suggests the tumor is also a symptom of a disturbed regulatory landscape. If so, then restoring the landscape might be as important as targeting the tumor. This does not abolish surgery, chemotherapy, or immunotherapy; it reframes them as one set of tools in a larger effort to restore collective control.

The body already attempts something like this on its own. Healthy epithelia (the sheets of cells lining organs and skin) can sometimes recognize and eliminate transformed neighbors through collective actions, as described by Fujita in 2015. One mechanism is apical extrusion: abnormal cells are squeezed out of the epithelial layer, like a defect popped from a fabric by surrounding threads. This kind of tissue-level tumor suppression is not a single gene acting heroically inside one cell; it is a neighborhood response. It depends on mechanical coordination, signaling, and an intact sense of what "belongs".

The fact that these collective defenses exist should change how we read early tumorigenesis. A cell may acquire a mutation today and never become cancer because it is eliminated, normalized, or kept quiescent by its environment. Even when early lesions appear, many remain dormant for years. The body's governance is not perfect, but it is persistent.

Bioelectric normalization belongs in this family of governance strategies. Why might voltage patterns have such leverage? One reason is that membrane potential is not just a messenger; it is a gatekeeper. Voltage influences the movement of ions such as calcium, which in turn controls gene expression, cytoskeletal organization, and proliferation. A sustained shift in membrane potential can bias the cell's entire regulatory network toward a more proliferative or more differentiated state. Because voltage patterns can couple across tissues—via gap junctions and other pathways—they offer a way to impose a coherent state over many cells at once. That coherence is exactly what a tumor tends to resist.

Another reason is conceptual. In earlier chapters, we treated intelligence in living systems as the ability to pursue goals across space and time, using feedback to correct errors. A tissue that maintains an organ's shape is solving a problem: it is keeping a pattern stable despite noise, injury, and turnover. Bioelectric signals are part of the tissue's internal measurement system—one of the ways it represents "where we are" relative to "where we should be." If those representations can be corrected, then behavior can be corrected. In that sense, bioelectric interventions resemble recalibrating a mis-set thermostat rather than replacing the entire heating system.

This is where the phrase "talking them back into the fold" becomes more than a slogan. The "talk" is a change in the informational environment. The "fold" is the tissue's attractor state—a stable pattern of cell behaviors that maintains an organ. One does not persuade a cell with arguments; one changes the constraints that make certain behaviors stable and others unstable.

It would be irresponsible to imply that such approaches are ready to replace established cancer treatments. Much of the work has been done in model organisms and developmental contexts where patterning signals are unusually strong and plasticity is high. Adult human tissues are different; tumors are genetically diverse and embedded in complex immune and stromal environments. Yet model systems are not toys. They are laboratories for principles. The principle here is that malignancy is sometimes reversible at the level of phenotype, and that long-range physiological signals—including bioelectric ones—can participate in that reversal.

Even in conventional oncology, the idea of normalization is not alien. Some therapies aim to normalize tumor blood vessels rather than destroy them, improving oxygenation and drug delivery. Differentiation therapies, such as those used in certain leukemias, attempt to push malignant cells into a mature state where they stop dividing uncontrollably. Immunotherapies seek not just to kill tumor cells directly but to restore an immune system's ability to recognize and constrain them. These are all, in different languages, attempts to rebuild governance.

What bioelectricity contributes is the suggestion that governance has an additional layer: a pattern-control layer that is neither purely genetic nor purely chemical, but distributed and field-like. That layer may be especially relevant for understanding why tumors can arise in regions of chronic injury or inflammation, where tissue-level signals are persistently altered. It also suggests new kinds of combination therapies: interventions that make the tissue environment more "legible" and coherent could, in principle, make other treatments more effective and reduce the selective pressure for aggressive escape strategies.

At a dinner party, one might summarize the whole argument like this. A tumor is not only a bad cell; it is a bad neighborhood. Not bad in the moral sense, but in the regulatory sense: a place where local feedback overwhelms global constraints. Sometimes the fastest way to change behavior is not to remove the actor but to change the stage directions that everyone is already following.

There is a final twist, and it is both hopeful and sobering. If malignancy can be suppressed by restoring collective signals, then cancer research is also, inevitably, research into the nature of the body's collective intelligence. What are the variables the tissue measures? What are the error signals that trigger correction? How far can the body plan reach in adult life, and where does it fail? These questions are not only medical. They reveal what kind of system a multicellular organism is: not a machine assembled from parts, but a society of agents kept in alignment by continuous information.

The chapter now stands at a natural threshold. We have seen defection as a breakdown of cooperation, and we have seen that the breakdown often involves a shrinking horizon of control. We have also seen hints that the horizon can be expanded again, by restoring the signals that knit cells into a coherent pattern. What remains is the most unsettling and practical question: how does a tissue decide, in real time, who belongs and who must be expelled? The next chapter turns from re-integration to surveillance and conflict—the everyday policing mechanisms by which collectives keep defectors rare, and the ways in which those mechanisms themselves can be turned, tragically, against the body.

Chapter 7

Interrogating the Alien

To distinguish a thinking system from a complex machine, scientists must act as interrogators, setting traps and puzzles that mere reflexes cannot solve. By subjecting cells and tissues to environments their ancestors never encountered, we reveal the boundary where mechanical reaction ends and true problem-solving begins.

7.1 A Turing Test for Tissues

Alan Turing's original provocation was not really about machines. It was about *standards.* If you want to ask whether something can think, Turing suggested, stop arguing about what the word *think* means and build a situation where a judge must decide, on the basis of performance, whether the hidden participant is human or not. The brilliance of the proposal was its stubborn practicality: it turned an unproductive metaphysical question into an experimental one.

The temptation, when we carry this idea into biology, is to imagine a parlor game with a petri dish. We picture a scientist asking questions and a tissue replying in Morse code. That's entertaining, but it misses the point. A *Turing-like* test for tissues is not a quiz show. It is an adversarial way of telling the difference between two things that can look identical at first glance: (1) a system that merely *runs through a script* when poked, and (2) a system that can *use information* to steer itself toward an outcome when the script is no longer enough.

The word we need is *novelty*. Reflexes can be exquisitely complicated and still be blind. The real diagnostic is what happens when you change the rules of the game in a way the system could not have been tuned for by its immediate history. If a pattern persists only in familiar circumstances, it may be the biological equivalent of a wind-up toy: beautifully engineered, profoundly limited. If the pattern survives into unfamiliar territory—and, even better, improves there—we start to suspect something like decision-making.

This section asks how to do that honestly. How do you interrogate a living collective—a sheet of cells, a developing organ, a regenerating fragment—without smuggling in your conclusions? How do you test for agency without assuming a brain? And how do you avoid being fooled by the most dangerous thing in the laboratory: your own talent for storytelling?

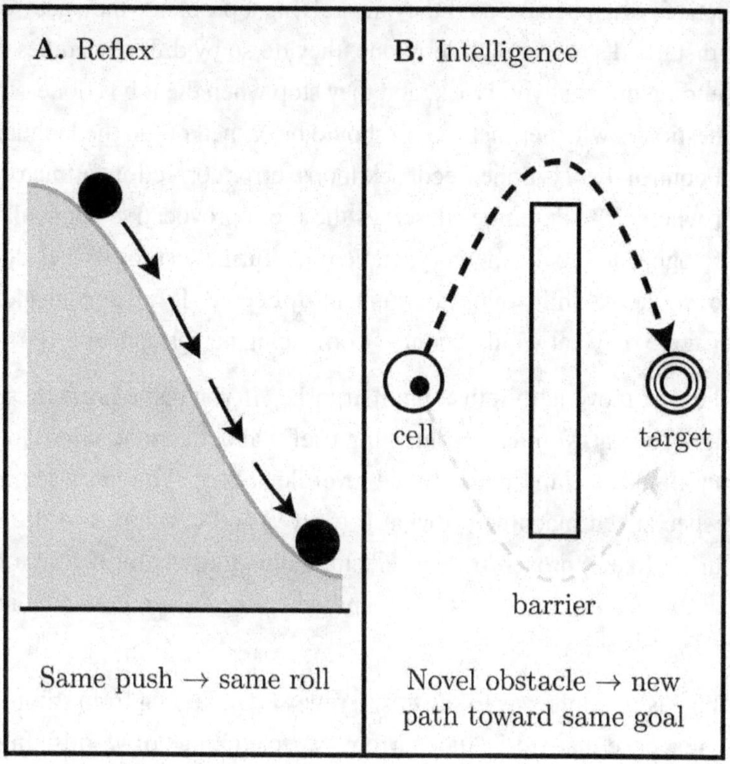

A diagram like this is deliberately unfair to gravity. A ball rolling downhill is not *stupid*; it is obedient to a law that applies everywhere, all the time. But that is exactly why it serves as a clean contrast. The ball's trajectory is explained once you know two things: the shape of the hill and the force pulling it down. Change the hill and the path changes, but only in the way a calculation changes when you alter its input. There is no need to imagine the ball *trying* to reach the valley. The endpoint is simply where the constraints place it.

Tissues, too, are constrained. They obey physics; they obey chemistry; they obey the limitations of their components. Yet living collectives often behave as if they are solving a problem: they get from a disturbed state to an orderly one, they do so by different routes depending on what you block, and they stop when the job is done. The question is whether that "as if" should be cashed out in the language of control theory alone (feedback loops, error correction, attractors), or whether it sometimes deserves the more provocative vocabulary of cognition (goals, memory, choice). A Turing test for tissues does not settle the philosophical argument by decree. It forces both sides to agree on what would count as convincing performance.

The first move is to refuse familiar tasks. If you test a heart by asking it to beat, you learn something useful about cardiac physiology, but almost nothing about tissue-level flexibility. You are watching a specialized machine do what it evolved and developed to do. A Turing-like approach tries to create a situation where the obvious, hardwired response would fail, and where success requires reorganizing behavior.

There is a simple way to say this: a reflex is a mapping from stimulus to response that works in a narrow, expected range of worlds. Intelligence is a mapping that remains useful when the world changes.

Of course, biology complicates this clean distinction immediately. Many reflexes are plastic; many plastic responses are reflex-like. Even bacteria can adjust their behavior based on gradients and history. The line between "pre-programmed" and "adaptive" is not a line at all but a spectrum, and living systems have had billions of years to migrate along it. That is why a tissue Turing test needs to be less like a single exam question and more like a battery of tricks, each designed to catch a different kind of impostor.

A historical anecdote helps here, because scientists have been trapped by impostors before. In the early twentieth century, the German horse known as Clever Hans appeared to do arithmetic, answer questions, and even tell time by tapping its hoof. People were not merely credulous; many were careful observers. The debunking, by psychologist Oskar Pfungst, did not rely on insulting anyone's intelligence. It relied on designing the right perturbation. Pfungst showed that Hans succeeded when he could see the questioner and failed when he could not. The horse was not doing math; it was exquisitely sensitive to involuntary human cues that signaled when to stop tapping. The lesson was not "horses are dumb." The lesson was that complex behavior can be explained by a narrow channel of information that the experimenter did not realize they were providing.

Tissues can be our Clever Hans. We can easily build mazes, gradients, wounds, and scaffolds that accidentally leak a hidden cue. We can mistake a physical bias for a decision. We can build an environment that practically drags the tissue to the answer and then applaud its "problem-solving." A tissue Turing test, if it is worth the name, must be designed the way Pfungst designed his trials: to remove unrecognized hints and to force the system to generalize rather than to exploit a loophole.

What does that look like at the bench?

Consider a cell migrating through a landscape. Cell motility is not mysterious magic; it is built from actin filaments, adhesion molecules, and signaling pathways that respond to chemical and mechanical cues. If you place a cell in a gradient of an attractant chemical, it tends to crawl up the gradient. That is chemotaxis, and it can be described with exquisite molecular detail. Yet chemotaxis alone is not a strong test of agency because it is a well-worn, evolutionarily familiar task. It is like asking a calculator to add.

Now change the situation. Put the attractant behind an obstacle that blocks direct access. The gradient may be distorted; the most obvious "follow the steepest slope" rule may lead straight into a dead end. In a strict reflex picture, the cell should stall or keep pressing into the barrier. In a more competent picture, it may explore laterally, sample alternatives, and eventually find a route around. The key is not that it moves around the obstacle once. The key is that it can do so across different obstacle shapes, different positions, and different kinds of blocked pathways, without you having tuned the system for each particular geometry.

This is where novelty becomes a litmus test. A one-off demonstration can be seductively cinematic: a time-lapse video of cells "deciding" to turn left makes for a wonderful seminar slide. But a Turing-like interrogation demands a hostile mindset: can the same system be made to fail by rearranging the environment in a way that defeats a simple rule? Can we generate families of problems that share a goal but differ in their surface details, and then ask whether performance transfers?

For tissues, the most interesting goals are often not about locomotion but about *form*. Developmental biology, traditionally, has been narrated as a kind of recipe: genes specify proteins; proteins set up gradients; gradients carve the embryo into parts. That story is powerful, but it can leave you with the impression that form is printed rather than achieved. And then, every so often, the embryo behaves in a way that sounds less like printing and more like proofreading.

Take a familiar household analogy. Imagine assembling a piece of furniture from a flat-pack kit. If the instructions are wrong, or a piece is missing, the assembly fails. That is what a purely feed-forward script looks like. Now imagine assembling the same furniture but with a team that keeps checking the final picture and correcting their work as they go. If a screw is missing, they improvise. If a panel is reversed, they notice and fix it. That is what feedback looks like. The astonishing thing about many developing and regenerating tissues is how much they resemble the second scenario.

A tissue Turing test tries to exploit this: it introduces defects that cannot be solved by the original, normal pathway, and then watches whether the tissue finds another route to a sensible outcome. In some organisms, if you cut away a part that would normally provide an instructive signal, development does not simply halt; it reroutes. In regeneration, if you injure tissue in a way that blocks a standard repair mechanism, repair can proceed through alternative programs. These are not miracles; they are the consequence of distributed systems with multiple overlapping ways to sense state and to change it. But they are also not easily reduced to a single reflex arc. They look like a kind of competence: the ability to reach a target state from many starting points.

One concrete example, by now famous because it feels like science fiction, comes from experiments in which clusters of frog embryonic skin cells were assembled into new shapes. In their usual context, those cells contribute to a tadpole's body. In a novel context—cut free, rearranged, and placed in a dish—they can form coherent little collectives that move, push particles, and interact with their environment in ways that were not part of their ancestral "job description." The drama of these constructs is not that they are alive; skin cells are alive in any context. The drama is that the collective behaves as if it has a *task*, and that its behavior depends on body plan in a way that makes it tempting to talk about a primitive kind of planning.

A skeptic can respond, reasonably, that this is just biomechanics: cilia beating, adhesion, friction. The point of a Turing-like test is not to silence that response but to sharpen the contest. If it is mere biomechanics, then changing the environment in certain ways should predictably break the behavior. If, instead, the system can be coaxed into achieving the same outcome by different microbehaviors under different constraints, then the "mere" begins to look like a sophisticated control policy enacted by tissue.

Designing such tests forces a practical question: what is the tissue's "conversation"? Turing's original imitation game used language because language is a rich medium in which humans display understanding. Tissues do not talk, but they do exchange signals: ions flowing through channels, electrical coupling through gap junctions, mechanical tension through cytoskeleton and extracellular matrix, molecular messages diffusing and being detected. If you want to interrogate a tissue fairly, you must speak in the tissue's idiom. That means setting up environments where the tissue can sense state and act on it, and where success requires integrating information across space and time.

One can phrase the core challenge as a problem of attribution. Suppose a tissue repairs a defect. Did it do so because it contains a fixed blueprint that, when damaged, automatically runs a repair subroutine? Or did it do so because it detected an error relative to some internal standard and acted to reduce it? Those two explanations can generate the same outward behavior in familiar situations. They diverge when you create circumstances the blueprint did not anticipate.

Here is a simple thought experiment. Imagine you have a developing structure that normally forms a left-right pair—say, two organs. You perturb the system so that one side begins to form in the wrong place. A feed-forward system should continue along its script, producing a misplaced organ. A regulative system might adjust positions, perhaps even moving tissue, until the relative arrangement becomes more normal. The difference is not in whether genes are involved (they always are) but in whether the process is sensitive to global error and can correct it.

Notice how close this is to how we test artificial systems. When engineers want to know whether a robot is robust, they do not ask it to walk across the same laboratory floor a thousand times. They spill sand, remove a leg, change the lighting. They are not being cruel; they are trying to discover whether the robot contains a general strategy or a brittle sequence of triggers. A Turing test for tissues is the same ethic applied to living matter: perturb and see whether the system's competence survives.

Because the word *intelligence* is so loaded, it helps to be explicit about what such a test is *not* claiming. Passing a tissue Turing test does not imply consciousness. It does not imply self-awareness. It does not imply that the tissue has beliefs. It implies something narrower and, in many ways, more scientifically valuable: that the system has enough internal state, and enough capacity to alter its behavior based on that state, that we can no longer model it as a simple stimulus-response device without losing explanatory power.

The most instructive moments in these tests are often the failures, because they reveal what the tissue's world looks like from the inside. Suppose a cell collective reliably finds its way around opaque barriers but fails when the barrier is porous to certain chemicals. That tells you which cues it uses. Suppose a regenerating fragment can restore missing structures unless a particular signaling pathway is blocked, in which case it settles into a stable but abnormal form. That tells you something about the space of attractors it can reach and the channels of communication that keep its parts coordinated. The interrogation is not a moral trial; it is a way of mapping the system's capabilities and its blind spots.

The problem, again, is Clever Hans. How do we avoid confusing an exploit with a competence?

One safeguard is randomization. If you want to test whether a tissue can solve a class of problems, you should generate many versions of the problem, ideally with parameters chosen randomly within constraints, and test performance across them. Another is *ablation of cues*: remove one potential information channel at a time. If the behavior survives the removal of a particular cue, the tissue was not relying on it exclusively. A third is *counterfactual environments*: create a world where a simple heuristic would predict the wrong move. For example, in navigation tasks, you can set up gradients that point away from the true target, forcing exploration or memory.

A more subtle safeguard is to attend to *time*. Reflexes can look clever in the short run. A thermostat is impressive the first time you see it stabilize a room's temperature; it is less impressive once you realize it cannot learn that you prefer it warmer in the evenings. Time gives you a way to test whether a tissue has a form of memory: does its behavior change with experience, and does that change persist through disturbance? Some organisms, like planarian flatworms, have made this question more than a philosophical tease. If an animal can be trained and then loses its head—and later regrows a head—what, exactly, is remembered? The details of that story belong later in this chapter, but the methodological lesson belongs here: a Turing-like interrogation does not merely ask whether a system can reach a target once. It asks whether the system can *improve*, and whether improvement can survive the kind of disruption that would erase a simple mechanical state.

At this point, an objection usually appears, wearing the sensible clothes of caution: isn't all this just rebranding? Isn't "goal-directedness" merely our description of a system that tends toward certain stable outcomes?

Sometimes, yes. There is no virtue in anthropomorphism for its own sake. But there is also a cost to refusing the language of goals when it is the simplest way to compress what we see. In engineering, we routinely describe systems as having goals without imagining that they daydream about them. A cruise control system has a setpoint. A thermostat has a target temperature. These phrases do not smuggle in consciousness; they acknowledge that the right level of explanation is not the microphysics of electrons but the macrobehavior of error correction.

The gamble of this book is that tissues may deserve an analogous treatment: not because they are tiny persons, but because they may be composed of parts that negotiate, compromise, and coordinate toward stable outcomes using internal signals that function like a distributed control network. Calling that "intelligence" is a provocation; calling it "competence" is often safer. Either way, a Turing-like test is the discipline that keeps the provocation honest.

One reason to value such tests is that they change what counts as a good scientific model. If you model a tissue as a blueprint executed step-by-step, your model might reproduce a normal developmental trajectory and still be worthless, because it will shatter under perturbation. If you model the tissue as an adaptive controller, your model must do more than replay the movie of development; it must keep the plot coherent when the script is vandalized. The model must *interact* with the world, not merely unfold.

This brings us back to Turing's deeper insight. The imitation game was never about finding a magical essence of thinking. It was about insisting that claims be tied to performance under interrogation. A tissue Turing test is a way of holding ourselves to the same standard when the "agent" is a sheet of cells. Instead of asking whether it *is* intelligent, we ask what it can do when we stop helping it, when we stop asking it to behave in the ways we expect, and when we confront it with problems that did not exist in its evolutionary past.

If that sounds harsh, it is also, strangely, respectful. It treats living matter as an opponent worthy of good questions.

The next step is to stop treating success and failure as binary outcomes and to begin measuring degrees of competence. Once you have a battery of interrogations, you can ask not only *whether* a tissue adapts, but *how much*, *how flexibly*, and over what range of time and space its "concern" extends. That shift—from a yes/no verdict to a calculus of agency—is where this story goes next.

7.2 The Calculus of Competence

After you have watched a tissue do something that looks, uncomfortably, like problem-solving, a familiar unease sets in. Perhaps it was only a trick of the environment. Perhaps we smuggled the answer in through an unnoticed cue. Perhaps our own hunger for meaning did the rest. The previous section argued for a method: confront living matter with novelty and see whether it merely reacts or genuinely adapts. Now comes the harder, less glamorous task. If novelty is the provocation, measurement is the discipline.

The difficulty is not that biology is messy (it is), but that we have inherited a set of categories that were built for animals with nervous systems. We are used to judging intelligence by speed, flexibility, memory, and planning, all measured through behaviour. With tissues, the most interesting behaviours may be invisible unless you know where to look. A developing face that quietly corrects an error over weeks is not a rat learning a maze. A sheet of cells that reroutes electrical signals after injury is not a bird learning a song. Yet there is still a question of agency hiding in both cases: how much can the system *control its own outcomes* in the face of disturbance?

That phrasing is deliberately plain. Agency, here, does not mean consciousness. It means the ability to keep a preferred state on track when the world pushes you off it. A thermostat has a kind of agency in this sense; it resists drift. A bicycle, by contrast, has no agency even though it can coast smoothly for a while. One system actively corrects error. The other merely persists until it cannot.

If this sounds like engineering, that is no accident. The language of control theory gives us tools that are oddly well suited for the very problem that makes "intelligence without brains" so contentious. Control theory was invented for artefacts, and it is largely indifferent to the substrate. It asks what variables a system monitors, what levers it can pull, how quickly it responds, and how robustly it reaches its setpoint when the world is uncooperative. Those are exactly the questions we need if we want to talk about competence without slipping into either mysticism (everything is alive, therefore everything thinks) or reductionism (it's all chemistry, therefore no one is allowed to use verbs like *want*).

A calculus of competence begins with a shift in what we count. Instead of asking, "Did the tissue succeed?" we ask, "How did success depend on pressure?" The pressure can be mechanical, chemical, electrical, or informational. You squeeze, block, scramble, distract, and then you measure how the system's performance degrades. Intelligence, on this view, is not a trophy you award; it is a curve you trace.

The metaphor of a curve is not just rhetorical. Imagine plotting performance on the vertical axis and difficulty on the horizontal axis. A brittle system looks fine until, suddenly, it fails catastrophically. A robust system degrades gracefully: it may slow down, become less precise, or choose a cheaper strategy, but it continues to make progress. Humans are like that. When deprived of sleep, we become sloppy but functional. When overwhelmed, we take shortcuts. Competence reveals itself not in perfection but in the shape of the fall.

One way to make this concrete is to borrow a concept that is intuitive in everyday life: *time horizon.* A time horizon is how far ahead a system behaves as if it cares. A thermostat has a tiny time horizon.

It corrects the temperature now and forgets it immediately. A chess player, even a mediocre one, has a longer time horizon. They sacrifice a pawn now to win a position later. In between lie countless biological systems. A single cell responding to a toxin has a short time horizon; it changes gene expression and membrane properties to survive the next minutes or hours. A developing tissue that repositions parts to achieve a viable anatomy has a longer time horizon; it is making corrections whose benefits appear days or weeks later.

There is also a *space horizon*: how large a patch of the world a system can effectively control. A bacterium has a small space horizon; it adjusts its own position. An organ has a larger one; it coordinates many cell types over centimetres. An animal, of course, can push that horizon outward through locomotion and tools. The important point is not the exact numbers but the direction: as you go from molecules to cells to tissues to organisms, the "cone of concern" expands in time and space. Something like intelligence begins to appear when a system can act now in order to change a distant future, or when it can influence not just its own immediate boundary but a wider territory of conditions.

These horizons already suggest why the binary question "Is a tissue intelligent?" is so unhelpful. Tissues often sit in the middle of the map. They are not as locally trapped as molecules, and not as globally strategic as animals. Their competence, when it exists, can be subtle: a kind of patient, distributed problem-solving in the space of anatomy.

How do you quantify this without turning every experiment into a philosophical referendum? You measure *trade-offs*. A reflex is cheap: it is fast and reliable in its expected world. A more agent-like system pays costs for flexibility. It must gather information, store some memory of what it found, and choose among alternatives. Those costs show up as time, energy use, variability, and sometimes even mistakes. In other words, competence is not only about success; it is also about the signature of *search.*

Search is what a ball rolling downhill does not do. It never samples alternatives. A competent tissue, by contrast, may try one route, then another. It may oscillate before settling. It may send exploratory protrusions or transient signals that do not directly contribute to the final solution but help map the constraints. If you can detect this kind of exploratory behaviour, you can begin to separate a simple feed-forward script from a closed-loop controller.

The challenge is that tissues do not always move in space. Much of their "search" happens in physiological space: patterns of voltage across membranes, patterns of gene expression, patterns of tension and adhesion. When those internal variables shift around before the system converges on a stable anatomy, you are watching a kind of internal deliberation. Calling it deliberation may be too human; calling it dynamics may be too empty. The calculus of competence tries to keep both possibilities in play by focusing on observable consequences: do internal states vary in ways that predict improved outcomes under novel constraints?

A concrete example makes the stakes clearer. In a study of developing frog tadpoles, researchers perturbed craniofacial structures—jaw and branchial arch elements that, in healthy animals, occupy specific positions and shapes. The surprising observation was not merely that defects occurred (development is sensitive), but that some perturbed structures moved over time toward more normal configurations, eventually becoming statistically difficult to distinguish from unperturbed controls. That kind of normalization invites a control-theoretic interpretation. Something in the developing system appears to be comparing "where this part is" to "where this part should be" and then adjusting growth and movement accordingly. It is as if a construction crew, having placed a doorway slightly off, notices the misalignment later and nudges the frame into place while the building is still malleable.

The reflex explanation would say: the tissue simply follows local rules, and those local rules happen, in typical conditions, to yield a normal face. The competence explanation says: the tissue can detect global error and correct it. These explanations can coexist in the same organism. Local rules can produce global patterns, and global feedback can ride on local mechanisms. The point of measurement is to discover how much of the outcome depends on a global "setpoint" versus an automatic cascade.

Here is where perturbation experiments become more than vandalism. If you introduce a defect and then block one of the plausible correction pathways, what happens? If the tissue still normalizes, it has alternative routes. If it fails, you have identified a critical channel. Either way, you can quantify competence by mapping the system's redundancy: how many independent ways it has to steer toward the target morphology. Redundancy is not a mere luxury in biology; it is often the signature of a system that must function in a noisy, damaged world.

A historical anecdote can help us appreciate why redundancy and feedback have always haunted explanations of life. In the nineteenth century, as embryology grew into a disciplined science, researchers argued fiercely about whether development was preformed (a tiny, complete organism unfolding) or epigenetic (a structure emerging through interactions). One famous figure in this debate was Hans Driesch, who, in the 1890s, separated early sea urchin embryos into individual cells. If development were a strict mosaic—each cell carrying a fixed piece of the final blueprint—the result should have been partial, monstrous fragments. Instead, he observed that separated cells could develop into smaller but relatively complete larvae. Driesch drew grand philosophical conclusions about a vital force, but the experimental fact survived the metaphysics: early embryos can regulate. They can compensate. They can, within limits, recover form when you scramble the parts.

Modern developmental biology has no need for Driesch's vitalism, but it has inherited his problem. How can a system be so forgiving? How can it produce a coherent whole when the initial conditions are altered? One answer is that development is not merely an execution of instructions but a feedback process: the system continually measures what has been built so far and adjusts what comes next. That is a recipe for robustness, and robustness is a measurable property.

So what are the metrics? A calculus of competence does not insist on a single number. It provides a family of measures, each aimed at a different aspect of agency.

One metric is *error correction capacity*: how large a deviation the system can tolerate while still reaching the target state. You can operationalize this by applying perturbations of increasing magnitude—larger wounds, stronger chemical blocks, more extreme rearrangements—and measuring the probability of recovery or the degree of normalization.

Another metric is *policy flexibility*: how many distinct trajectories can lead to the same outcome. A brittle system has one way to succeed. A flexible system has several. You can estimate this by tracking the internal and external paths taken across replicate experiments. If individuals reach the same endpoint through different intermediate states, you have evidence of multiple compensatory strategies.

A third metric, and one that becomes particularly interesting for tissues, is *persuadability*. Persuadability asks whether you can change what the system "aims" for by changing the incentives and constraints. The term is provocative on purpose. We are used to persuading animals: you offer reward, you apply punishment, you change the environment so that one behaviour becomes cheaper than another. With tissues, persuasion does not mean bribery; it means altering the landscape of signals so that the system settles into a different stable state.

A simple example comes from bioelectricity. Cells maintain differences in voltage across their membranes, and groups of cells can form patterned voltage distributions. Those patterns are not mere by-products; they can influence gene expression and, in some cases, anatomical outcomes. If you can alter ion channel activity or gap junction connectivity and thereby push a tissue toward forming an organ in an unusual place, you have demonstrated a striking kind of persuadability: the tissue's "default" goal state can be re-specified by altering its internal communication. From the perspective of competence, this matters because it suggests that the system is not a rigid machine with a single destiny. It is a controller whose setpoints can be shifted.

Persuadability also gives a way to compare systems along a continuum. A simple reflex is hard to persuade; it always does the same thing when triggered. A more agent-like system can be nudged: it may trade one objective for another, or accept a compromise when resources are scarce. In animal behaviour, this is obvious. A hungry animal will take more risks. A tired animal will settle for a smaller reward. In tissues, analogous trade-offs may appear as changes in growth, movement, or patterning when you alter the costs of different pathways. A tissue that can reroute repair around a blocked pathway is displaying flexibility. A tissue that can be driven into a different stable form by changing the "reward" landscape of signals is displaying something closer to preference.

One might worry that this is simply relabeling physics. But the relabeling earns its keep if it improves prediction. Persuadability is not poetry; it is a way to design experiments. If you think a tissue is running a fixed developmental script, then changing certain boundary conditions should not systematically redirect the endpoint; it should merely cause failure. If you think the tissue is a controller with adjustable setpoints, then changing those boundary conditions should redirect outcomes in structured ways. That is a falsifiable difference.

All of this can be made more rigorous by borrowing from information theory. A competent system must, in some sense, use information from the environment to reduce uncertainty about what action to take. You can quantify how much a perturbation increases uncertainty about outcomes, and how much the system reduces that uncertainty through its responses. A system that behaves the same way regardless of context is low-information; it is not listening. A system whose behaviour depends sensitively on relevant features of the environment is high-information; it is discriminating.

The strongest version of this idea is to treat the tissue and its environment as an interactive game. You, the experimenter, choose perturbations; the tissue responds. Over many rounds, you learn the tissue's "policy"—not as a metaphor, but as a statistical mapping from sensed states to actions. The more compactly you can describe that mapping while still predicting behaviour across novel tasks, the more confident you can be that you have captured something real. This is also where the adversarial spirit of a Turing-like test returns: a good model should not merely replay known outcomes, but withstand new interrogations posed by someone trying to make it fail.

This is not merely academic. If tissues possess measurable competence, then regenerative medicine becomes less like carpentry and more like diplomacy. The traditional engineering approach to repair is to build the missing part and install it. But living systems are not inert structures; they have their own tendencies, their own internal coordination. If you want a limb to regrow or a wound to heal without scarring, you may not need to micromanage every cell. You may need to set the right goals and open the right channels of communication so the tissue can do what it does best: solve the problem of getting back to a stable, functional form.

At the same time, competence has a darker implication. If tissues can pursue setpoints, they can also pursue the wrong ones. Cancer can be seen, in part, as a breakdown of the multicellular agreement about what counts as a proper goal state. A cell collective that once cooperated to maintain an organ's architecture becomes a population pursuing local growth advantages. In that light, measuring competence is inseparable from measuring *alignment*: whose goals are being optimized, and at what scale? Even without invoking moral language, the biological question is sharp. When do local objectives (cell survival, proliferation) conflict with global objectives (organ function, organism survival), and what control mechanisms normally keep them in harmony?

The calculus of competence therefore points toward an even more practical question: what organisms, what tissues, and what experimental systems are best suited for this interrogation? Not all living matter is equally legible. Some systems regenerate spectacularly; others scar. Some embryos regulate robustly; others are fragile. Some tissues offer convenient handles for manipulating bioelectric signals; others resist. To do this science, researchers need partners that are generous with their tricks.

Those partners include immortal flatworms that can rebuild themselves from fragments, and frog embryos whose cells can be rearranged, persuaded, and watched as they construct surprising forms. They are, in a sense, our best "aliens": living systems that do not behave the way our nervous-system-centred intuitions predict, and that therefore force us to refine our tests.

The next section turns to those organisms—flatworms, frogs, and the engineered collectives assembled from their cells—not as curiosities, but as instruments. They are the places where the abstractions of time horizons, space horizons, error correction, and persuadability become visible enough to measure, and strange enough to be worth arguing about.

7.3 Flatworms, Frogs, and Frankenstein

A theory about intelligence without brains can stay elegant for exactly as long as it avoids the laboratory. The moment you ask for evidence, you need creatures that will tolerate your questions. Not all living systems are generous in this way. Many animals are too complex to interpret and too ethically fraught to manipulate freely; many cells are too simple to exhibit the kind of large-scale competence that makes the debate interesting. The field that studies problem-solving in tissues therefore has an unusually intimate relationship with its model organisms. It depends on a few biological "accomplices" that are hardy, legible, and just strange enough to force us to revise our instincts.

Three have become emblematic. Planarian flatworms, which treat dismemberment as an inconvenience. Frog embryos, whose developing bodies expose their internal coordination in real time. And the engineered "Frankenstein" collectives sometimes called xenobots, assembled from ordinary cells into shapes evolution never tried. Each is a different kind of interrogation device. Together, they let us ask whether the competencies we measure—error correction, flexibility, persuadability—are quirks of particular species or signatures of something deeper about living matter.

The emotional tug of these systems is not incidental. They unsettle us because they scramble categories we use to stay comfortable. We like minds to be located in brains, and we like bodies to be obedient to genomes. A flatworm that can learn and then rebuild its head makes memory feel less securely housed. A tadpole that repairs a misbuilt face makes development feel less like a printout. A clump of skin cells that cooperates to move and manipulate the world makes individuality feel negotiable. The point is not to replace biology with metaphors, but to use these organisms to identify which metaphors survive contact with measurement.

Planaria are the most disarming of the three, partly because they look like nothing at all: a thin, soft ribbon gliding on a film of mucus, with two dark eye spots that give it a faintly cartoonish expressiveness. Their claim to fame is regeneration. Cut a planarian into pieces and, within limits, each piece can rebuild a whole worm. It is an old story, so old that it risks becoming a party trick. But the deeper lesson is not that planaria regenerate; it is that they regenerate *reliably*. They do not merely grow tissue. They restore a coherent anatomy: head at one end, tail at the other, organs in the right places, proportions that make sense for the size of the fragment. That reliability is exactly what a calculus of competence tries to quantify. If there is an internal anatomical "goal state," planaria behave as though they can find it from many different starting points.

The history here is surprisingly long and, in its early chapters, surprisingly philosophical. In the late nineteenth and early twentieth centuries, regeneration was one of the arenas in which biologists fought about whether life required special forces. If an organism could reorganize itself after violent perturbation, perhaps that meant there was an immaterial plan guiding it. That conclusion now feels like a relic, but the experimental puzzle remains. When a planarian fragment decides where to put a new head, what information is it using? It is not reading a blueprint from the outside. It is using internal signals, distributed across cells, to infer what is missing and to coordinate growth and patterning.

What makes planaria especially valuable for SUTI research is that they allow an interrogation that would be grotesque in most animals: you can remove the head, including the brain, and the animal will often regrow it. That turns a philosophical question into a technical one: if the animal can be trained, and then you remove the brain, does anything of the training persist?

The claim that learning can survive decapitation sounds like a horror story until you hear it as an experimental design. The logic resembles the Turing-like tests we discussed earlier. If a behaviour is a brittle reflex localized to a particular structure, destroying that structure should destroy the behaviour. If some memory or bias survives, then either (a) the memory was stored outside the brain, or (b) the brain, when regrown, can recover the memory from non-neural traces. Either possibility stretches the usual mapping between memory and neurons.

To make such tests credible, researchers have used training paradigms that avoid wishful interpretation: automated assays, objective scoring, and controls that distinguish genuine learning from fatigue or sensory damage. Planaria can be exposed repeatedly to a stimulus pattern that, over time, produces a measurable change in behaviour. When such trained animals are decapitated and allowed to regrow heads, they can show a "savings" effect: they relearn faster than naive animals, as reported by Shomrat and Levin in 2013. The careful reader will feel two impulses at once. One is excitement: perhaps memory is more widely distributed than we thought. The other is scepticism: perhaps some subtle peripheral change makes the behaviour easier to acquire the second time. Both impulses are appropriate. The value of the planarian system is precisely that it keeps the question experimentally open. It does not force a mystical answer, but it forces us to admit that "brain equals memory" is not a law of nature; it is a hypothesis that can fail in particular ways.

In the language of competence, planaria offer a long time horizon and a generous error correction capacity. They treat major anatomical disturbance not as the end of the story but as the beginning of a repair problem. If you want to study how tissues coordinate across space, you want an organism that routinely solves a spatial coordination problem: rebuild a head with the right polarity and proportions. If you want to study how stable anatomical targets are maintained, you want an organism that can return to those targets repeatedly.

Frog embryos, particularly those of *Xenopus* species, provide a different kind of generosity. They do not regenerate like planaria, but they are unusually accessible windows into development. Their embryos are large, robust, and amenable to manipulation. You can alter ion channel activity, change mechanical constraints, perturb signalling pathways, and then watch how the body plan unfolds. If planaria are the masters of repair, frog embryos are the masters of construction.

Construction is where the blueprint metaphor has been most seductive. It is tempting to think of development as a genetic program executed step by step, like a well-written recipe. Yet anyone who has cooked knows that recipes are not enough. You taste, you adjust, you compensate for humidity and heat and the temperament of ingredients. A purely feed-forward recipe works only in a world without surprises. Development does not have that luxury. Embryos develop in changing environments, with noisy molecular events and occasional damage. The fact that they usually reach a viable form suggests that they, too, have something like a tasting spoon: feedback.

The craniofacial normalization phenomenon in tadpoles is a powerful example because it is so intuitively human. A face is not merely tissue; it is a geometry. When jaw elements are perturbed—shifted, malformed, displaced—some of them can drift over time toward more typical positions and relationships, as quantified by Vandenberg, Adams, and Levin in 2012. The details are technical, involving careful morphometric measurements rather than dramatic photographs, but the conceptual punch is simple: the system does not just grow; it corrects. It behaves less like a printer and more like a sculptor who keeps stepping back to check proportions.

The most interesting questions begin when you ask what the embryo is checking against. Is there a fixed "image" of the correct face, encoded genetically? Or is correctness defined relationally, as a set of constraints among parts (this must connect to that, this must align with that), discovered and maintained through local interactions? The second possibility is particularly compatible with a distributed intelligence view. In many complex systems, global order emerges from local rules plus feedback. A termite mound does not require an architect termite; it requires termites that can sense local conditions and respond in ways that stabilize certain macroscopic patterns. The embryo may be doing something analogous, except with cells negotiating through chemical gradients, mechanical tensions, and bioelectric signals.

Bioelectricity, in this context, is not the dramatic crackle of neurons firing. It is the quieter fact that all cells maintain electrical states: differences in voltage across their membranes, maintained by ion channels and pumps, and shared through gap junctions that connect neighbouring cells. These electrical states can form patterns across tissues. Crucially, manipulating those patterns can change developmental outcomes. Experiments have shown that altering membrane voltage patterns in frog embryos can produce startling respecifications, including the formation of eye structures in locations where eyes do not normally appear, such as the tail or back, as Pai and Levin demonstrated in 2012. Whatever else one thinks about the interpretation, this establishes something important for a calculus of competence: tissues are *persuadable* through a non-neural control layer. You can change what the tissue builds by changing how cells communicate electrically.

That matters for two reasons. First, it suggests that gene expression is not the only "language" of development; there are higher-level coordination channels that can override or redirect gene-driven tendencies. Second, it offers an experimental handle. If you want to interrogate tissue-level decision-making, you need knobs you can turn. Ion channels and gap junctions are such knobs. They allow you to perturb the internal information network of a developing system and see whether it compensates, reroutes, or collapses.

There is a social dimension to why *Xenopus* has become such a central character in this story. It is not only that the embryos are convenient. It is that the community has built tools, norms, and shared knowledge around them. A model organism is a kind of collective agreement: we will all ask questions in this dialect so that our answers accumulate. In that sense, frog embryos are not merely biological material; they are part of a cultural infrastructure for interrogating life.

Then there are the "Frankenstein" constructs, which earn their nickname by breaking the quiet pact we have with natural organisms. Planaria and tadpoles are products of evolution. However strange they seem, they are still within the repertoire of what life has tried before. Xenobots, by contrast, are assembled, first demonstrated by Kriegman, Blackiston, Levin, and Bongard in 2020 and extended in 2021. Researchers take embryonic cells—often from *Xenopus*—and arrange them into novel geometries. The cells are not genetically engineered in the usual sense; they are ordinary cells placed into unusual contexts. The surprise is that the collective can exhibit coherent behaviour: self-directed motion, interaction with objects, and, under specific conditions, the ability to corral free cells into piles that then develop into offspring-like collectives capable of repeating the process for limited rounds. The word *replication* here is not metaphorical, but it is also not the familiar replication of genes or dividing cells. It is a new mode of reproduction based on movement and assembly.

Why should this belong in a chapter about interrogating an alien? Because xenobots let you do something that is nearly impossible with natural organisms: you can create a large set of body plans that differ only in geometry, and then measure how that geometry changes the competencies the collective expresses. If a planarian regenerates, you can marvel and then struggle to disentangle what is specific to planaria. If a xenobot moves, you can ask what parts of the behaviour are generic properties of ciliated cells and what parts depend on the collective's architecture. You can treat the living system almost like a new material whose properties depend on shape and coupling.

This is where the link to the previous section's metrics becomes vivid. Xenobots are a platform for measuring policy flexibility and error correction in a controlled way. You can put them in different environments, change obstacles, alter available free cells, and see whether the collective finds alternative strategies. You can test persuadability directly by changing the medium or by altering cell-cell coupling. You can quantify time horizons by asking how long the collective maintains a behaviour in the absence of immediate reinforcement. You can quantify space horizons by measuring how large an area it can effectively manipulate relative to its size.

Perhaps the most conceptually important feature of xenobots is that they force a separation between *genetic history* and *behavioural possibility*. A common intuition says: if a system does something, evolution must have selected that something. Xenobots challenge this. Their behaviours arise in contexts that have never existed in nature, at least not in the specific engineered forms. This does not mean evolution is irrelevant. The cells come with evolved machinery: motility, adhesion, sensing, and communication. But the behaviour is not a direct product of selection for that behaviour. It is a product of what the cellular machinery can do when organized differently. In other words, xenobots reveal latent competencies in tissues—capabilities that evolution did not need to express in that particular way, but that were available nonetheless.

That is precisely what an "interrogation" aims to uncover. When you place cells into a novel task environment, you are not merely testing their reflexes. You are mapping the space of possible problem-solving strategies that their internal machinery supports. Xenobots make that mapping unusually explicit because the experimenter (and often an algorithm) proposes the body plans, and the cells supply the rest. The resulting behaviour feels alien because it is the product of a negotiation between design and biology, between imposed geometry and intrinsic cellular tendencies.

There is, inevitably, a cultural echo in calling these constructs Frankenstein-like. Mary Shelley's novel was not really about monsters; it was about responsibility and misunderstanding. Victor Frankenstein creates life and then refuses to understand it, and the tragedy follows from that refusal. The modern scientific version is gentler but still morally charged. When we assemble living collectives into new forms, we are not only discovering what tissues can do; we are learning what it means to intervene in systems that have their own internal coherence. Even if xenobots are far from sentient, they are not inert. They are closer to organisms than to machines, and that in-between status demands care in how we speak about them and how we use them.

All three models—flatworms, frogs, and engineered collectives—therefore serve as different answers to a single practical question: how do you make the "alien" of tissue intelligence talk back? Planaria answer by being nearly impossible to permanently break, turning amputation into a reversible perturbation. Frogs answer by developing in a way that exposes coordination and allows powerful physiological interventions. Xenobots answer by letting us propose forms and then watch what living matter does with them, revealing competencies that are not tied to a natural life history.

Each also keeps us honest in a different way. Planaria remind us that memory and identity may not reside where we assume. Frogs remind us that development can be regulative and that bioelectricity can act as a control layer. Xenobots remind us that the same cells, in a new geometry, can express unexpected goals, and that "what life does" is not exhausted by what life has already done in nature.

The next challenge is to bring these systems under a unified explanatory umbrella. It is one thing to collect marvels. It is another to identify the common control principles that make them possible: how signals integrate across tissues, how setpoints are represented, how errors are detected, how competing objectives are resolved. Having met the protagonists, we can now ask what kind of internal conversation allows a fragment to know it needs a head, a tadpole to nudge a face toward normality, or a clump of cells to behave as a coherent agent. The alien has given us enough clues to begin reconstructing its language.

Chapter 8

Building New Minds

Biology is commonly thought of as a fixed catalog of species, but recent breakthroughs in synthetic biology suggest life is far more malleable than previously believed. By liberating cells from their usual anatomical contexts, we are discovering that tissues possess a latent, programmable intelligence capable of building organisms evolution never imagined.

8.1 The Xenobot Paradox

Biology has a habit of hiding its most unsettling talents in plain sight. We grow up thinking of skin as the least mysterious organ: a wrapping, a barrier, a living raincoat. It stretches, heals, and ages, but it does not *do* very much. And yet, when a small amount of frog embryonic tissue is removed from the story it was meant to tell, something remarkable happens. The same cells that would ordinarily help make the surface of a tadpole can be coaxed into making a tiny, coherent creature-like thing that moves, navigates, and sometimes cooperates with its peers, as shown by Kriegman, Blackiston, Levin, and Bongard in 2020 and 2021. No brain. No nerves. No special genetic tricks. Just a rearrangement of familiar living material, and a shift in the boundary conditions under which that material is allowed to act.

The paradox is not that we can build something new from biology. People have done that for decades. The paradox is that we did *not* have to invent the key behaviors. We did not have to design the motion the way we design a motor, or design the control system the way we design a circuit. We did not even have to edit the genome, the canonical place where modern imagination stores the instructions for what bodies can become. The key behaviors were already latent in the cells. What changed was the context: the goals and constraints imposed by the embryo were removed, and the cells were allowed to settle into a different kind of collective.

To appreciate why this is so strange, it helps to begin with a very ordinary scene in developmental biology. A frog embryo is a bustling construction site. Cells are dividing, migrating, sticking to each other, pulling on each other, and sorting themselves into tissues with distinct mechanical properties. Under a microscope, this can look almost serene: a smooth choreography of rounds and layers. But the serenity is deceptive. Every move is the result of negotiation. Cells exchange signals, read tension in their neighbors, and respond to chemical gradients; they take local cues and contribute to global shape. A normal embryo is not simply a pile of parts that assembles; it is a distributed system that makes decisions.

Now imagine a researcher performing a small, careful act of violence: a *scraping*, as it is sometimes described in laboratory shorthand. A set of embryonic cells is removed from its native location. The embryo, if still alive, will attempt to compensate. The removed cells, now in a dish, are suddenly homeless. They have their molecular tools, their adhesive molecules, their ability to contract and relax. They have their genetic identity as frog cells. What they lack is the embryo's agenda: the larger anatomical project that normally tells them, in effect, *you are skin, and here is the map of where skin goes.*

At first glance, one might expect the cells to do what many isolated tissues do: form an amorphous clump and gradually decay. Instead, under the right conditions, they do something closer to improvisation. They assemble into a compact form, often a roughly spherical aggregate. They seal their surfaces, establishing a coherent boundary. They become, in a limited but striking sense, an individual.

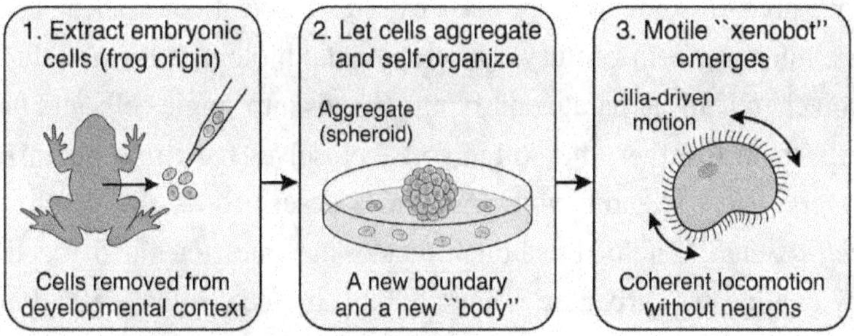

The name that eventually attached itself to these constructs, *xenobots*, is revealing. The prefix *xeno-* means foreign or strange, and it is not only the laboratory method that feels strange. It is the fact that the finished entity is built from perfectly ordinary biological stuff. This is not a cyborg. There is no silicon. There is no motor. There is not even a wire. What appears in the dish is, in a literal sense, frog tissue.

Yet it behaves like a little machine.

The first temptation is to say: *well, the cells are alive, so of course they move.* But embryos are full of living cells that do not crawl away. In the frog embryo, skin cells end up as a sheet, a tilework that protects and senses. In the dish, the same kinds of cells can form a compact body and propel it across a surface. The relevant distinction is not life versus non-life; it is *task context.* Cells carry a repertoire of possible behaviors, and that repertoire is selectively expressed depending on the architecture they find themselves in.

One can make this feel less mystical by remembering a precedent that long predates xenobots: the classic tissue dissociation experiments of the mid-twentieth century. Developmental biologists learned that if you chemically separate embryonic tissues into single cells and then mix them together, they often sort themselves back into organized layers. This phenomenon, sometimes described as *self-sorting* or *self-assembly*, helped establish that tissues are not glued together by one master carpenter. They are built by local rules: cells prefer certain neighbors, they pull with characteristic forces, and large-scale order emerges. It was an early lesson in the intelligence of collectives, though it was not called that at the time.

Those older experiments, however, mostly concerned *anatomical correctness*: could cells rebuild something resembling an embryo? Xenobots forced a different question: could cells, given the chance, build something that resembles *a device*?

Part of the answer lies in a piece of biological machinery that rarely gets top billing in popular accounts of life: *cilia.* Cilia are tiny hair-like structures that protrude from cells. In many contexts, they act like oars. In your own body, cilia help clear mucus from airways. In ponds, ciliated microorganisms use them to swim. In an embryo, cilia can appear on epithelia (sheet-like tissues) and drive flows.

When embryonic frog cells are assembled into a small mass, some of them differentiate in ways that include making cilia on their outer surface. If those cilia beat in coordinated fashion, the whole aggregate can move. Nothing about this requires neurons. Coordinated beating can arise through local coupling and the physics of fluids near a surface. The "controller," if we insist on the term, is partly in the tissue's patterning dynamics and partly in the geometry of the body it happens to occupy.

This is one reason the xenobot story is not simply a tale of a clever laboratory trick. It is a window into a deeper point: much of what we attribute to centralized control is, in living systems, outsourced to form and material. In a conventional robot, a body is passive hardware. In biology, the body is an active participant in problem-solving. It senses forces, it reacts, it stabilizes. A sheet of cells can close a wound without consulting a brain. A cluster can round itself into a sphere because surface tension and adhesion make that state energetically favorable. A ciliated surface can generate motion because the molecular machinery of cilia, once deployed, produces periodic force as naturally as a heart produces a beat.

The *accidental* aspect of the discovery is important because it reminds us how rarely we think to ask certain questions. Many scientific breakthroughs begin not with a grand prediction but with someone noticing that the control experiment is doing something odd. Here, the oddity was a kind of competence: tissue that should have been merely tissue was acting like an agent.

Agency is a slippery word, and it tends to summon philosophy. For our purposes, it can be kept practical: an agent is a system that can steer outcomes into a narrower range than chance would allow. A rock does not steer; it rolls wherever gravity and friction dictate. A bacterium, though brainless, biases its movement toward nutrients. A xenobot is not a frog and not a machine in the usual sense, but it can still display a crude form of outcome control: it moves persistently, it can push against tiny obstacles, and under some conditions it can contribute to collective behaviors. It is a system that can, within limits, *do*.

If you are tempted to object that this is all metaphor, consider how routinely we accept similar language elsewhere. We say that immune cells *patrol*, that tumors *invade*, that bacteria *strategize*. We do so because these are not random motions; they are coordinated, feedback-driven behaviors that achieve functional ends. Xenobots deserve the same descriptive generosity, if only because the alternative is to pretend that coherent behavior without a brain is impossible, which biology has never agreed to.

Still, the unsettling part is not that cells can move. It is that they can be recruited into an unfamiliar kind of whole. A frog embryo is a familiar whole: it has a developmental plan shaped by evolution, and it produces an organism that can survive in a pond. A xenobot has no evolutionary history as a xenobot. No ancestors lived as a living sphere in a dish. And yet the cells do not simply flail. They coordinate.

What could coordination mean here? At minimum, it means that the cells manage their mutual tensions and adhesions so that the aggregate does not disintegrate. It means that some cells take on roles that make a functional surface. It means that the system maintains a boundary between inside and outside. Those are not trivial achievements for a mass of cells thrown together.

There is a revealing way to think about this: the genome is not a blueprint for one body; it is a library of competencies that can be expressed under different conditions. The same frog genome that can build a tadpole can also, when its cells are liberated and recombined, build other coherent structures. This does not mean the genome is infinitely flexible. It means that development is not a single hard-coded program but a set of interactive processes that can settle into multiple stable outcomes. Some of those outcomes are the ones evolution has repeatedly used. Others are available but seldom visited.

A historical anecdote helps anchor this in the broader arc of science. In the nineteenth century, when early microscopists first watched cells divide, it was tempting to imagine that a developing organism was like a pre-written text being read out loud. By the early twentieth century, with the rise of embryology, another image gained ground: the organism as a sculpture, carved by forces and gradients. The modern picture is stranger still: the organism as a society. Cells communicate, form alliances, and enforce boundaries. They follow local rules, but the collective exhibits a kind of goal-directedness: it tends toward certain anatomical endpoints and repairs deviations when perturbed. Xenobots are a small, vivid demonstration that this "society" does not dissolve the moment you remove the "government" of the nervous system. The capacity to coordinate is distributed far more widely than our intuitions allow.

One might worry that calling this coordination "intelligence" dilutes the word. But the dilution concern only makes sense if intelligence is defined as whatever brains do. The entire point of this book is that brains are not the only place where problem-solving happens. The xenobot paradox forces a clean separation between intelligence and consciousness. There is no reason to think a xenobot has experiences. There is, however, good reason to think it has capacities that resemble problem-solving in the engineer's sense: robustness under disturbance, exploitation of physical dynamics, and reliable production of behavior from minimal instruction.

The minimal instruction, in this case, is largely *shape*. A tiny change in geometry can turn a drifting spheroid into a purposeful mover. This is where the story begins to feel less like magic and more like design space. The cells bring the actuators and sensors, but geometry channels those capabilities into different behavioral regimes. It is the biological analog of a sail: the sail is not a motor, yet it transforms wind into directed motion. Give the same wind to a different shape, and you get a different trajectory.

Once you accept shape as instruction, another unsettling thought follows. If ordinary embryonic cells can be assembled into a new kind of coherent agent, then the line between "organism" and "artifact" is not a wall but a gradient. That gradient has been there all along. A beaver dam is an artifact produced by an organism. A coral reef is an architecture produced by a collective of organisms. Xenobots are stranger only because the artifact is made *of* the organism, and because the designer is a human rearranging parts that evolution never placed in that configuration.

At this point, a reader often asks a reasonable question: if there is no genetic modification, what makes xenobots "new"? The answer is that novelty in biology is not confined to new genes. It can arise from new couplings: new spatial arrangements, new mechanical constraints, new environmental contexts. Put the same cell types into a different arrangement and you have, in effect, a new body plan on a tiny scale. This is a kind of innovation that development can express immediately, without waiting for mutation and selection. Evolution, of course, is the long-term editor of what persistently works in the wild. But development is the short-term generator of coherent forms given a genome and a context.

The dish, in this sense, is not merely a container. It is a new world with new rules: smooth surfaces, controlled fluids, absence of predators, abundant oxygen. The embryo is a world too, with its own constraints. By moving cells between worlds, researchers reveal what those cells are capable of in principle, not merely what they do in their usual home.

There is also a deeper lesson about error correction. In an embryo, if you remove a portion of tissue, development often compensates. The final organism may be smaller or altered, but it is frequently still recognizable. That implies that embryos do not simply "unfold" a pre-specified structure; they regulate toward it. Xenobots, though simpler, show a hint of the same regulatory tendency: aggregates can maintain coherence, and under some circumstances even heal small damage. This matters because it shifts our picture of living matter from fragile to resilient. Engineers spend enormous effort building systems that fail gracefully. Biology often gets this property for free, because its components are adaptive.

The xenobot paradox, then, is not a party trick of frog cells. It is a proof of concept that the basic units of life are not merely parts but agents in their own right, able to form new coalitions and enact new behaviors when placed under new constraints. If that feels like a big claim to rest on a millimeter-scale blob in a dish, remember that big claims in science often begin with small creatures. The fruit fly taught us genetics. Sea urchins taught us fertilization. Bacteria taught us the logic of regulation. The xenobot teaches something about the plasticity of agency.

One can end, as popular accounts sometimes do, with a flourish about "living robots." But the more fruitful ending is quieter. Xenobots are not the arrival of machines that happen to be alive; they are the revelation that living tissue is already a kind of machine, in the sense that it can be harnessed, redirected, and made to solve problems without explicit instruction. Once you have seen that, it becomes difficult to look at any body the same way. Your own tissues are not passive matter awaiting commands from your brain. They are collectives of cells that negotiate, repair, and regulate, most of the time without you noticing.

And this naturally raises the next question. If a cluster of cells can compute with its body simply by being shaped in the right way, then perhaps the distinction between "robotics" and "biology" is less about materials and more about where the problem-solving happens. A metal robot is built to think in silicon and act with motors. A biological construct can think, in a limited but real sense, with its morphology and its self-organizing dynamics. The next section follows this thought to its logical conclusion: what does it mean to build *robots made of meat*, and what advantages—perhaps even what dangers—come from treating living tissue as a new kind of programmable hardware?

8.2 Robots Made of Meat

Once you have seen a xenobot scoot across a dish, the word *robot* starts behaving badly. It becomes less a category than a provocation. A robot, in the everyday sense, is a manufactured thing: gears, code, batteries, perhaps a polite voice. A frog cell is not manufactured; it is grown. It is not coded by a programmer; it is regulated by chemistry. Yet the xenobot looks like something built *for* a purpose. It does not merely *happen*; it *acts*.

The unease is useful. It forces a question we usually avoid by habit: what exactly distinguishes a machine from an organism? If the answer is "machines are designed and organisms are evolved," xenobots immediately muddy the water, because they are designed from evolved parts. If the answer is "machines are predictable and organisms are messy," xenobots again interfere, because living tissue can be surprisingly reliable when its own self-organizing tendencies are doing the hard work. If the answer is "machines follow external control while organisms control themselves," then living constructs sit right on the seam, because their control is internal, but their *goals* are often set by an external designer.

The phrase that helps make sense of this seam is *morphological computation.* It sounds like jargon, but it names a simple idea: the shape and physical properties of a body can perform part of the "computation" that we usually imagine happening in a brain or a microprocessor. Here "computation" does not mean arithmetic. It means turning messy inputs (forces, flows, collisions) into useful outputs (stability, movement, sorting) without a central controller doing explicit calculations.

A familiar example lives in your hand. Pick up an object you have never held before—a new mug, a tool from a friend's kitchen drawer. You do not solve a set of equations to determine how much to squeeze at each point. Your fingers deform. Skin friction, tendon elasticity, and the geometry of joints do much of the work. The nervous system contributes, certainly, but it is not issuing micromanaged commands to every muscle fiber at every millisecond. Your hand is a soft robot whose body is doing part of the thinking.

Roboticists have been learning this lesson for a long time, sometimes grudgingly. Early robots, in factories, could afford to be stiff and simple because the world they lived in was controlled: identical parts, fixed locations, fenced-off humans. As soon as you ask a robot to do something like "walk on uneven ground" or "pick up a strawberry without crushing it," stiffness becomes a liability. The robot needs a body that can absorb uncertainty. That push toward softness is not a mere engineering fashion; it is an admission that the world is too rich to be handled by centralized computation alone.

Biology did not have to admit this. Biology never pretended otherwise. Animal bodies are built from compliant materials that can store and release energy, distribute stresses, and self-correct small errors. A tendon behaves like a spring. Cartilage compresses and rebounds. The very *sloppiness* that engineers usually dislike is, in living systems, part of the strategy. A system that can bend without breaking does not need to predict every perturbation. It can let the perturbation happen and then recover.

Xenobots sharpen this idea because they strip away the usual suspects. When a dog runs, you can always say: "it's the brain." When a xenobot moves, there is no nervous system issuing commands. Motion is generated by tissue properties that, in a frog embryo, were never "meant" to make a free-living body. The controlling intelligence is distributed across the material itself. In other words, the xenobot is not a robot that *has* a body; it is a robot that *is* a body.

That is the first blurring: hardware and software begin to merge. In conventional machines, the distinction is clean. The hardware is the physical substrate. The software is the abstract set of instructions that can, in principle, be copied onto different hardware. In living systems, the instructions are inseparable from the substrate. A change in tissue stiffness can change behavior as surely as a change in code. A change in geometry can transform a chaotic twitch into directed locomotion. The "program" is partly chemical, partly electrical, partly mechanical, and it is distributed.

There is a second blurring, one that matters outside the lab: *repair*. We accept that machines need maintenance, and we treat repair as an external intervention. A car does not heal; it is healed by a mechanic. Living tissue, by contrast, treats damage as a signal. A wound triggers cascades: clotting, inflammation, cell migration, rebuilding. Some xenobots, when cut or damaged, can reconstitute some of their structure, at least enough to keep functioning. In robotics terms, this is astonishing. It is as if a drone that loses a propeller could reorganize its remaining parts and continue flying.

Repair changes the economics of complexity. In traditional engineering, the more complex a machine becomes, the more failure modes it has, and the more fragile it often is. Biology pays a different price.

Living systems are complex, but they are also continuously renewing. Cells are replaced. Proteins are turned over. The body is never quite the same object from one day to the next. The "robot made of meat" is not maintained from outside; it maintains itself from within.

If this sounds like science fiction, it is worth remembering that the fantasy of self-maintaining machines has been with us for a long time, and not only in literature. A concrete historical anecdote comes from the mid-twentieth century, when scientists and mathematicians began thinking seriously about self-reproducing and self-repairing automata. John von Neumann, a founder of modern computing, worried about a deep constraint: ordinary machines require a factory to build them. They cannot copy themselves without an entire infrastructure. He asked whether one could design a system that, in the right environment, could assemble a copy of itself from available parts. His answer, developed in abstract models, was that self-reproduction is possible in principle if the system contains both a description and a constructor that can interpret it. The point here is not the details of von Neumann's model. The point is that self-maintenance and self-replication were recognized as profound engineering problems, precisely because the usual machine paradigm does not include them.

Living tissue solved these problems long before engineers named them. Not perfectly, not without constraints, but robustly enough that multicellular organisms exist at all. Xenobots bring that solution into the engineering domain without importing the whole animal. They are a kind of minimal case: a body that can do something useful because its material contains built-in competencies for cohesion, patterning, and sometimes repair.

At this stage, it is tempting to say that the only difference between a biological robot and a conventional one is the material: carbon versus silicon, wet versus dry. But that is too shallow. The deeper difference is the *unit of design*. In conventional robotics, the designer specifies components and then specifies control policies: how sensors feed into decisions, how motors respond. In living robotics, the designer can often specify only a boundary condition—a shape, an arrangement—and then rely on the material's own dynamics to fill in the details. You are not writing a script; you are setting a stage and letting the actors improvise within constraints.

This is where morphological computation stops being a philosophical slogan and becomes an engineering strategy. Consider a simple mechanical analogy. A passive dynamic walker is a device that can "walk" down a slope with no motors and no active control, just carefully arranged legs and joints. Gravity supplies energy, and the geometry of the limbs turns that energy into a stable gait. The walker computes its next step through its own dynamics. It is not intelligent in the everyday sense, but it achieves a behavior we associate with agents, and it does so by embodying the solution in its morphology.

Xenobots are a biological cousin of that idea. Their movement is not planned; it is enacted by the collective behavior of cilia and tissue mechanics. Their "controller" is not a centralized logic unit; it is a set of coupled oscillators, adhesions, and feedback loops that happen to produce coherent motion when arranged appropriately. The difference is that the biological body can also *change itself*. Cilia can appear or disappear as cells differentiate. Tissue stiffness can shift with cellular state. A biological robot is not merely a cleverly shaped object; it is an object whose shape and capabilities can be actively maintained by living processes.

Once we accept this, the question "is it a robot?" becomes less interesting than "what does the robot metaphor buy us, and what does it hide?" The robot metaphor buys us a certain kind of clarity: we can ask what task is being performed, what constraints make it feasible, what design choices improve performance. But it hides the fact that living systems have their own priorities. Cells are not passive building blocks. They are agents with homeostatic drives: they tend to maintain their internal states, adhere to certain neighbors, avoid certain stresses, and respond to damage. When we assemble them into a new body, we are not merely arranging parts; we are negotiating with a population of microscopic entities that have their own local goals.

This negotiation helps explain why xenobots feel so different from genetically engineered organisms. Genetic engineering intervenes in the internal rules of the cell: it changes what proteins are made, what signals are sent, what circuits exist. Xenobot construction, in its most striking form, intervenes in the *ecology* of the cells: who touches whom, what boundary exists, what shape emerges. It is closer to architecture than to gene editing. If you want to change how people behave, you can change their biology (drugs, hormones), or you can change the layout of the city (streets, lighting, meeting places). Xenobots show how far the second approach can go.

Morphological computation also clarifies why "robots made of meat" can be energy-efficient. Traditional robots spend energy not only moving but also computing, sensing, and correcting errors. They must continuously measure and adjust because their bodies do not naturally stabilize. A living tissue, by contrast, can exploit physical relaxation toward stable states. It can use local chemical gradients as free signals. It can outsource parts of control to mechanics and to collective dynamics. This does not make it magical; it makes it economical.

The blurring line, however, is not only scientific. It is social. Words like *robot* carry cultural baggage: fear of runaway technology, fascination with artificial life, anxiety about what counts as "natural." Calling a xenobot a robot invites one set of reactions; calling it an organism invites another. The reactions matter because they shape governance and funding and public trust. A living construct that is biodegradable and confined to a lab dish is, in practical terms, less threatening than many technologies already in circulation. Yet the word *robot* can trigger a sense of uncontrollability. Meanwhile, the word *organism* can trigger worries about moral status: if it is alive, does it deserve protection? If it is designed, are we "playing God"? These concerns are not always well calibrated to the actual system in question, but they cannot simply be dismissed. They are part of how societies decide what kinds of research to support.

Here morphological computation offers an unexpected ethical angle. If a great deal of "control" in living constructs is achieved by shaping tissue and exploiting its natural dynamics, then the designer's hand may be less direct than in conventional engineering. You do not specify every action; you shape a system that tends to produce certain actions. This can be reassuring, because it suggests constraints and predictability rooted in physics and biology. It can also be unsettling, because it means the system may display emergent behaviors that were not explicitly planned. Emergence is not the same as danger, but it does complicate the question of responsibility. When a body computes through its own dynamics, the design space becomes wide, and surprises become normal.

A concrete example helps. Imagine you want a xenobot to move through a cluttered environment. In a conventional robot, you would add sensors, write code for obstacle avoidance, test, and iterate. In a living construct, you might instead adjust the body shape so that collisions naturally reorient it toward open paths, much as a Roomba bounces into a workable coverage pattern without mapping your home. The "algorithm" is partly the geometry. You can achieve a behavior that looks purposeful without installing a purpose-representing module. That is morphological computation in action, and it is exactly why the line between "alive" and "engineered" becomes hard to draw.

At a dinner party, someone will eventually ask the blunt question: "So are these things *thinking*?" The best answer is to separate two senses of thinking. If thinking means having an inner life, a point of view, an experience of the world, then there is no reason to attribute that to a simple living construct built from frog cells. If thinking means solving a problem—maintaining cohesion, producing directed movement, repairing damage, coordinating with neighbors—then yes, in the limited sense that matters to engineers and to this book. Thinking, in that sense, can be a property of tissue.

This is not a rhetorical trick. It is a shift of scale. We are used to putting "mind" in the head and "mere biology" everywhere else. But when you look closely, tissues are full of feedback loops. They measure and respond. They correct. They maintain boundaries and repair breaches. In an embryo, these loops guide anatomy; in a dish, they can guide behavior. Xenobots are one dramatic case, but the underlying principle is general: life computes with matter.

The most practical consequence of this principle is that it opens a new design philosophy. Instead of treating the body as a passive platform for a controller, we can treat the body as a partner in control. Instead of writing ever more elaborate code to handle every contingency, we can sculpt systems whose physical form naturally handles contingencies. In biological materials, that partnership extends even further: the body can grow, heal, and adapt. A robot made of meat may be less precise than a robot made of metal, but it may be more robust in the face of the world's unpredictability.

This brings us to the edge of a larger landscape. If shape can act as instruction, then there is not just one way for tissue to become a functional agent. There are many. Some will resemble known organisms; others will not. The xenobot is a small point in a vast design space of possible forms, most of which evolution has never sampled because it had no reason to. To move from a few surprising constructs to a systematic understanding, we need a map—a way to think about all the bodies that could exist, not only the ones that do.

That map is the subject of the next section: the idea of *morphospace*, a conceptual landscape of possible forms. Once you start taking "robots made of meat" seriously, you inevitably start asking where, in that landscape, nature has already walked, where it has never gone, and what new kinds of minds might appear if we learn how to steer tissues through the unexplored regions.

8.3 Mapping the Morphospace

Once you accept that bodies can compute—that shape is not just a container for intelligence but one of its engines—a new kind of question becomes hard to resist. If a small clump of frog cells can be nudged into a new, functional form by rearrangement alone, how many other forms are waiting in the wings? How many ways can living tissue be organized into something coherent, capable, perhaps even useful, that evolution has never tried?

It is tempting to answer with a slogan: "infinitely many." But biology is never infinite in any practical sense. Bodies must obey physics. They must be buildable by development. They must function in some environment. So the real problem is subtler and more interesting. The real problem is to map the space of possible forms in a way that lets us see both the freedom and the constraints—and to understand why life, in all its exuberance, occupies only a narrow region of what seems possible.

This is where the notion of *morphospace* comes in. A morphospace is a conceptual map of form: an abstract space in which each point corresponds to a particular shape, architecture, or anatomical configuration. The trick is that a morphospace is not a photograph album of creatures we already know. It is a generator. You specify a set of parameters—numbers that can be varied—and those parameters produce shapes. The parameters might describe how tightly a shell coils, how long a limb is relative to a body, how frequently a branching structure splits, how thick a tissue layer becomes, or how stiff a segment is. You then allow those parameters to vary over their possible ranges, and you look at the resulting landscape. Some regions look familiar: they resemble real organisms. Other regions look bizarre, even impossible. The key word is *look*: the morphospace reveals what is geometrically possible in the model, which may be more permissive than what biology can actually build.

The value of this approach is not that it tells you what exists. It tells you what *could* exist, given a particular set of rules. And once you have that "could," you can place the "does" onto the map and ask why the living world clusters where it does.

A concrete historical example shows how powerful this can be. In the 1960s, the paleontologist David Raup became fascinated by a seemingly modest question: why do snail shells look the way they do? Shells vary enormously, yet the variation is oddly structured. You do not see every conceivable spiral. Raup's move was to treat shell shape as a mathematical consequence of a few simple parameters: how fast the shell expands as it coils, how far the generating curve sits from the coiling axis, and how quickly it moves along that axis. Change the parameters and you can generate a dizzying range of shells: tight coils, loose coils, tall spires, flat disks, forms that overlap themselves, forms that produce huge voids, forms that look like they belong in an alien aquarium.

Then Raup did something beautifully simple. He plotted real shells on this theoretical map. They did not fill the space. They occupied a restricted region. There were large areas of "possible" shell shapes that no known mollusc used. Raup's point was not that evolution had been unimaginative. His point was that empty regions of morphospace are often evidence of constraints. Some shapes may be mechanically weak. Some may be hard to build developmentally. Some may create apertures that cannot function for a living animal. Some may be perfectly viable but simply never reached by the particular historical paths available to molluscs. The emptiness itself becomes a clue.

A morphospace, in other words, is a way to turn absence into data.

It is easy, from a distance, to tell a comforting story in which evolution explores possibilities and keeps the good ones. But the morphospace lens forces you to notice that evolution explores possibilities in a very particular way. It is not a tourist wandering freely across a map. It is more like a process moving through a maze whose walls are built by development, physics, ecology, and history. Some paths are blocked. Some are narrow. Some lead to dead ends. And some open into huge rooms that, for contingent reasons, no one has entered yet.

This matters for our purposes because the previous sections established a key point: development is not a brittle script. It is a set of self-organizing processes that can settle into different outcomes when constraints change. Xenobots are tiny proof that new outcomes can appear without new genes. That makes morphospace not just a tool for paleontology, but a tool for engineering. If forms can be generated, searched, and instantiated, then mapping morphospace becomes a practical problem: it tells you where to look for new kinds of functional bodies.

Before we push too far, a warning is in order. Not all morphospaces are created equal. Some are *theoretical* morphospaces, like Raup's shells, where parameters generate shapes according to a model. Others are *morphometric* spaces, built from measurements of real organisms: you take a collection of shapes, quantify them (often in high dimensions), and represent the variation statistically. These different spaces answer different questions. A theoretical morphospace is good for asking "what does the model permit?" A morphometric space is good for asking "how do real organisms vary?" Confusing the two can lead to misleading conclusions, because a theoretical morphospace may include shapes that are mathematically definable but biologically meaningless, while a morphometric space may reflect only what has already evolved and thus hides the larger possibility space.

Even within theoretical morphospaces, the choice of parameters is not neutral. If you choose parameters that correspond to what nature already varies easily, your morphospace will make nature look adventurous. If you choose parameters that correspond to what nature finds difficult, your morphospace will be filled with empty regions, and you may be tempted to overinterpret them. The map is not the territory; it is an instrument. Like any instrument, it shapes what you can see.

And yet, even with these caveats, morphospace offers a disciplined way to talk about "possible creatures" without drifting into fantasy. It lets you ask questions that are otherwise too vague. How many degrees of freedom does a body plan have? Which variations change function a little, and which flip it into a new regime? Which shapes are stable under perturbation? Which shapes make movement easier? Which shapes make repair easier? Which shapes make cooperation possible?

The most important conceptual shift is that morphospace is not only about *appearance*. It is about *behavior*, because behavior is often a property of form interacting with environment. Think again of morphological computation. A body shape can turn random collisions into a stable orientation. A stiffness profile can turn periodic muscle contractions into efficient locomotion. A branching pattern can turn a simple pumping motion into widespread distribution of nutrients. When you vary form, you vary the computation performed by the body.

This is where the xenobot story and the morphospace idea lock together. In xenobot research, one can treat the design problem as a search through a morphology space: different geometries are tried (sometimes in simulation), evaluated for performance on a task, and then the most promising shapes are built out of living tissue. The task might be "move forward," "push a payload," or "corral loose cells into a pile." The point is not the particular task. The point is that the search is not primarily through genes; it is through *forms*. Biology supplies a material that can realize a chosen form and then self-organize the details. Computation supplies a way to explore vast numbers of candidate shapes without building each one by hand.

Here a useful analogy comes from a much older technology: shipbuilding. A skilled shipwright does not compute fluid dynamics for every plank. Traditional hull designs evolved through experience, and they were constrained by materials and tools. Modern naval architecture, by contrast, explores a design space computationally, testing hull shapes in simulation and wind tunnels, optimizing for stability and efficiency. The ocean has not changed. What has changed is our ability to search a large space of forms quickly. Xenobots represent a similar transition, but in living matter. The "ocean" is the physics of tissue and fluid. The "hull shapes" are configurations of cells. The novelty is that the material is itself adaptive: it does not merely hold shape; it participates in maintaining and expressing it.

There is also a deeper biological resonance. Evolution, too, is a search algorithm operating in morphospace. It proposes variations and filters them by survival and reproduction. But evolution's proposals are not arbitrary. They are generated by development. Development is the machinery that turns genotype into phenotype, and its biases shape which regions of morphospace are easily reachable. Some changes are easy: lengthen a limb, thicken a shell. Others are hard: rearrange the entire organization of tissues without lethal side effects. Developmental constraints carve grooves in morphospace. Evolution tends to slide along grooves, not because it lacks imagination, but because the machinery that produces variation is itself structured.

One of the most emotionally striking implications is that the creatures we know may be less a catalogue of what is possible and more a record of what was historically reachable. The vertebrate body plan, for example, is wildly successful. It is also oddly conservative: one head, one spine, two pairs of limbs (when present), the same broad arrangement repeated and modified. It is as if evolution found a powerful design early and then elaborated it endlessly rather than reinventing it from scratch. The morphospace view suggests that this conservatism may not reflect a lack of better options in principle. It may reflect the difficulty of crossing certain developmental distances.

Humans have a special relationship to this conservatism, because our medical ambitions constantly run into it. When a human limb is lost, we do not regrow it, not because the atoms are missing, but because the developmental programs that could rebuild a limb are not active in adult humans in the way they are in some other animals. The form "adult human with regrown arm" sits in morphospace as a plausible configuration of tissues, but it is not reachable by our normal regulatory pathways. If we could understand the control layers that steer tissue growth and patterning, we might be able to move an adult body into a region of morphospace that is currently out of reach. In that light, morphospace is not only about fantastical creatures. It is about unrealized medical outcomes.

The idea also changes how we interpret "impossible." In everyday language, impossible often means "I can't imagine how." In science, impossible should mean "incompatible with known constraints." Morphospace helps separate these. Some forms are impossible because they violate physics: an organism cannot be built out of tissue that has the tensile strength of smoke. Some forms are impossible because they violate developmental logic: the intermediate steps required to get there are lethal. Some forms are merely "unrealized": no lineage stumbled into them, or the environment never rewarded them, or a constraint that once mattered no longer does. The last category is where engineering becomes interesting, because engineering can change constraints. A dish can remove predators. A scaffold can provide support. A laboratory can supply energy sources that nature rarely offers. By changing the environment, we can make viable forms that nature never selected.

Here is where the book's theme of intelligence without brains returns in a new guise. If we think of intelligence as the ability to steer outcomes, then morphospace is a map of steerable outcomes at the level of anatomy. Developmental systems already display a kind of intelligence: they correct perturbations and converge on functional forms. What we are learning to do is to *guide* that intelligence, to set new targets, to alter the landscape so that tissues can settle into different attractors—different stable anatomical solutions. Xenobots are one small example: cell collectives that, under new constraints, settle into motile forms. But the broader ambition is to treat morphospace as a playground for new kinds of body-level problem solving.

This suggests a new way to frame the relationship between evolution, development, and engineering. Evolution explored morphospace slowly, constrained by reproductive success and historical contingencies. Development explores a much smaller region repeatedly, reliably building the forms that have worked before, while retaining latent capacities for self-organization and repair. Engineering, especially when paired with computation, can explore morphospace differently: rapidly, purposefully, and with new constraints. It can take advantage of the same self-organizing competencies that evolution harnessed, but it can aim them at tasks evolution never cared about: targeted drug delivery, tissue repair, microscale cleanup, or simply the scientific goal of understanding what living matter can do.

There is a final twist, and it brings us back to the emotional charge of the subject. The moment you begin mapping morphospace, you begin to see that familiar creatures are not the pinnacle of a ladder but the inhabitants of a neighborhood. We are used to thinking of "higher" and "lower" organisms, as if evolution were a contest with a single finish line. Morphospace dissolves that ladder. It shows a landscape with many peaks and valleys, many viable strategies, many weird corners. The frog is not an inevitable solution. It is one solution among many, stabilized by history. That is humbling in the best way.

At the same time, it is empowering. If life's forms are points in a vast space, and if tissues have generic capacities for self-organization, then building new minds becomes less like inventing from scratch and more like navigating. The critical question becomes: what are the coordinates? What parameters matter? How do we move reliably from one region to another without breaking the system?

Those questions do not end with xenobots. They broaden into the central challenge of this entire subject: understanding the control layers by which cell collectives decide what to build. Once you start thinking in morphospace, you begin to ask not only what forms exist, but what signals and feedback loops *select* among forms during development and regeneration. The next chapter turns from the geometry of possible bodies to the mechanisms that steer bodies through that geometry: how tissues store information, how they communicate it, and how life achieves its most impressive trick of all—building coherent selves, again and again, out of cells that have never been told the whole plan.

Chapter 9

The Software of Life

If biological hardware is not the only place intelligence resides, then medicine must evolve from molecular micromanagement to anatomical goal-setting. By learning the language of bio-electricity and cellular cooperation, we stand on the brink of triggering regeneration on command, forcing us to confront the profound ethical responsibilities of rewriting the software of living systems.

9.1 The Diplomat's Medicine

A bone breaks. A surgeon sets it, pins it, screws it, and waits for the body to do what bodies have done for hundreds of millions of years: knit mineral back together. A heart valve fails. Engineers cut and sew a new one into place. A nerve is severed. Surgeons try to reconnect it, millimetre by millimetre, and hope the axons find their way back. Modern medicine is astonishingly good at *mechanics*: replacing parts, patching holes, removing tumours, bypassing blockages. The underlying picture is often a kind of biological carpentry. When the structure is wrong, we impose the right structure by hand.

But there is another way to think about repair, one that feels less like carpentry and more like diplomacy. A diplomat does not move battalions like chess pieces. They alter incentives, set constraints, send messages, and let thousands of independent agents reconfigure themselves into a new stable arrangement. The diplomat's art is indirect power: the smallest intervention that causes a large system to reorganize its own behaviour.

Living tissue, at the scales that matter for healing and development, is exactly such a system. Cells are not inert bricks. Each cell is a tiny, busy creature: sensing, comparing, negotiating, cooperating, and sometimes defecting. If you tell a cell to divide, it does not merely divide; it divides *somewhere*, with *some polarity*, into a neighbourhood of other cells that have their own plans. If you tell a cell to become muscle, it becomes muscle in a tissue that must connect, align, and carry load. Shape is not painted on; it is the consequence of millions of local decisions that somehow add up to a leg, a kidney, or a face.

The obvious response is to micromanage. Turn the right genes on. Add the right growth factors. Seed the right scaffold with the right cell types in the right order. This approach has produced real successes, and it will continue to do so. Yet it also runs into a practical wall: the body is not a static design assembled once. It is a dynamic pattern that must be maintained against noise, injury, and time. The problem is not merely building tissue; it is getting tissue to *keep* building the right tissue, and to stop at the right moment, and to integrate with what is already there.

Diplomatic medicine begins from a different premise. Instead of asking, "How do we place the cells where we want them?" it asks, "What does the tissue collectively *think* it is supposed to be?" Not in a mystical sense, but in a control-theoretic sense: what internal signals define the target the tissue will work toward, and what messages can nudge that target?

To see why this matters, consider a classic, almost unsettling set of observations from the early days of experimental embryology. In the 1920s, the German embryologist Hans Spemann and his student

Hilde Mangold performed what became known as the "organizer" experiment. They transplanted a small patch of tissue from one amphibian embryo into another at a critical stage. The result was not a small local change but a wholesale reorganization: a second body axis began to form. In some cases you got what looked like a two-headed tadpole, an embryo persuaded to lay down a new blueprint. The transplanted tissue did not itself grow into a second animal like a grafted limb; rather, it *instructed* surrounding cells to participate in building an additional structure.

What made that experiment enduring was not the oddity of a two-headed amphibian. It was the implication that form can be redirected by relatively small informational interventions. A patch of tissue acted like a command, not a component.

That is a useful contrast to keep in mind when we talk about the more modern claim that there is a "software" layer of life. Software, in the everyday sense, is not the physical circuitry. It is the flexible pattern of states and signals that tells hardware what to do next. In biological hardware, the "circuits" are cells and tissues, with genes and proteins as essential components. Yet tissues also rely on signals that are neither genetic sequences nor soluble chemical gradients. They include mechanical stresses, electrical states across cell membranes, and networks of cell-to-cell communication that couple many individual decisions into one coordinated outcome.

The diplomatic metaphor becomes most vivid when we focus on one particular set of signals: bioelectricity.

Every living cell is electrically polarized. It maintains a voltage difference across its membrane, produced by ion channels and pumps that move charged atoms (like sodium, potassium, chloride, and calcium) in and out. Neurons use rapid electrical spikes to communicate, so we associate electricity with brains. But most cells are electrically active in a quieter way. Their resting voltage can change slowly, persist for long periods, and influence what genes are expressed, how cells migrate, when they divide, and how they adhere to neighbours. Cells can also connect to each other through gap junctions, tiny channels that directly couple their interiors. A tissue linked by gap junctions is not merely a collection of individuals; it is an electrical community. Ions and small signalling molecules can spread, allowing a pattern of voltage to exist across a sheet of cells the way a pressure pattern can exist across a weather map.

What does that buy you? Something that medicine has always wanted: a handle on *collective behaviour*.

Imagine trying to regenerate a limb by addressing each cell separately. You would need to specify cell types, positions, orientations, connectivity, growth rates, and termination conditions. You would also need to adapt as the system changes, because development is not a scripted assembly line; it is a feedback-driven process. Diplomatic medicine looks for a different lever: can you trigger a high-level instruction—"build a limb here"—and let the tissue solve the implementation details using its own adaptive machinery?

In nature, some animals seem to do exactly that. A salamander can regenerate an amputated limb with a fidelity that makes human surgeons jealous. It is not assembling the limb cell by cell using an engineer's blueprint; it is activating a program that recruits nearby cells, rewinds them into a more flexible state, coordinates their proliferation and patterning, and then stops when the limb is complete. The remarkable part is not only the regrowth; it is the *error correction.* Cut the stump at different angles, remove tissue, perturb the process, and the system often still converges on the correct anatomy. This is what engineers would call robustness: the ability to reach a stable goal from many starting conditions.

That kind of robustness is one of the strongest clues that there is an internal representation of the target morphology, a memory of what "complete" looks like. In a purely mechanical view, injury destroys structure and repair is local. In a goal-directed view, injury is a deviation from a setpoint, and regeneration is the system's attempt to return to that setpoint.

The diplomatic question is whether we can talk to that setpoint.

A concrete illustration comes from work on planarian flatworms, famous for their regenerative powers. Cut a planarian into pieces and many pieces can regrow into complete worms, complete with brains and eyes. More startlingly, experiments have shown that by perturbing the worm's physiological signalling—including its bioelectric coupling—you can sometimes induce stable changes in what it regenerates. The animal's cells do not merely regrow *something*; they regrow what they now "expect" the body plan to be. Under some conditions, a worm fragment can regenerate with altered polarity or with extra heads, and in some cases that altered regenerative outcome can persist across subsequent amputations even after the original perturbation is gone. The tissue behaves as if a memory has been edited: not the genome, but the patterning state that guides rebuilding.

If that sounds like science fiction, it is worth pausing to appreciate what is *not* being claimed. No one is saying a flatworm has a human-like mental image of itself. The claim is more modest and, in a sense, more radical: that tissues can occupy stable physiological states that bias them toward certain large-scale anatomical outcomes, and that those states can sometimes be switched. In dynamical systems language, you can picture development and regeneration as a landscape of attractors. A marble rolling on that landscape ends up in a valley. The valleys are stable outcomes: a one-headed worm, a two-headed worm, and so on. If you change the shape of the landscape—by altering connectivity, ion channel activity, or other network properties—

you can make a different valley more likely. The tissue then does what it always does: it rolls downhill, using its own rules, its own local problem-solving, toward the new stable form.

The diplomatic power is not in forcing each step, but in reshaping the space of possibilities.

This suggests a kind of medical intervention that is almost the opposite of surgery. Instead of cutting away diseased tissue and replacing it with engineered material, you might deliver a pattern of electrical stimulation, or a drug that modulates ion channels, or a biomaterial scaffold that changes the tissue's electrical microenvironment. The goal would not be to build the missing structure directly, but to *recruit the tissue's own construction crew* by changing the messages it is receiving.

There is an intuitive everyday analogy. When a city suffers a blackout, you could send technicians to flip every switch in every building to restore the pattern of light. That would be hopeless. Instead, you restore power to a few key substations and the city's internal wiring does the rest. Diplomacy is about finding the substations of morphogenesis.

The shift in mindset has practical consequences. A mechanic's medicine is comfortable with detailed control: if you can see it, you can fix it. Diplomat's medicine needs a different kind of knowledge. You must understand what signals coordinate many cells at once, what patterns act as "permissions" or "constraints," and how a tissue interprets those patterns.

One reason bioelectricity is so attractive here is that electrical states naturally integrate across space. Chemical signals diffuse and degrade; they are excellent for local gradients but harder to maintain as stable, tissue-wide patterns without constant input. Electrical coupling through gap junctions can produce coherent domains: regions of tissue sharing similar membrane voltage, separated by boundaries. Those boundaries can behave like functional borders, telling cells on one side that they are part of one structure and cells on the other side that they are part of another. The details are complex, but the guiding idea is simple: voltage patterns can act as a coordinate system, a kind of physiological map that helps cells decide what role they should play.

The diplomat's approach also reframes what counts as a successful therapy. In a purely mechanical repair, success is immediate: the graft takes, the stitch holds, the device functions. In an informational repair, success may look slow and slightly mysterious. The intervention is brief, but the outcome unfolds over days or weeks as the tissue reorganizes itself. And because you are steering a living collective rather than assembling a machine, variability is inevitable. Outcomes may be probabilistic. A given trigger might bias the tissue toward regeneration rather than scarring, or toward one anatomical pattern rather than another, without guaranteeing it every time. That is not necessarily a weakness; it is the signature of working with an adaptive system rather than against it. Still, it demands a new relationship between physician and patient: one that tolerates uncertainty, monitors trajectories, and adjusts signals rather than performing a single definitive fix.

There is also a quieter emotional shift in the metaphor. A mechanic tends to treat the body as passive matter. A diplomat must respect agency at the level being addressed. Again, not human agency, but control over outcomes. A tissue that can correct errors, converge on a stable form, and respond to perturbations has a kind of minimal agency: it can act to reduce the gap between its current state and a target state. Diplomatic medicine is the art of negotiating with that agency. You do not simply impose a design; you propose a new goal and provide the conditions under which the tissue can pursue it.

One can already see the outlines of how this could change regenerative medicine. Consider chronic wounds, where tissue fails to close properly and cycles through inflammation. Or spinal cord injuries, where scarring and inhibitory cues prevent regrowth. Or congenital defects, where early patterning errors propagate into anatomy. These are not merely missing parts; they are mis-specified trajectories. The classic intervention is to patch the trajectory locally: debride, graft, implant. The diplomatic intervention would be to identify the physiological "settings" that keep the tissue stuck—electrical, chemical, mechanical—and reset them so that the tissue's own behaviour shifts into a healing attractor rather than a pathological one.

It is tempting to see this as a universal solution, a master key. It is not. Bioelectric signals are one layer in a dense stack of interacting controls: gene regulation, biochemical gradients, immune signals, mechanics, metabolism, microbiome interactions, and more. A diplomat who knows only one language will not succeed in every negotiation. Still, the promise is not that electricity replaces genetics or chemistry. The promise is that there exist higher-level control inputs that can efficiently recruit many downstream mechanisms at once, and that bioelectricity may be one such input because it naturally couples cells into functional collectives.

At this point, it helps to return to the opening contrast: micromanagement versus prompting. In everyday computing, you can write a program that specifies every pixel, or you can call a library function: `draw_circle(x,y,r)`. The library call is a promise that thousands of operations will happen correctly without your supervision. The programmer trusts the subroutine because it has been tested and because it embodies a robust solution to a recurring problem.

Diplomatic medicine is a search for the body's subroutines. "Close this wound." "Build a branching network here." "Regrow a fingertip." "Restore left-right polarity." The audacity of the idea is not that cells are programmable; it is that they may already have rich built-in programs, and we simply lack the interface.

A historical anecdote from surgery makes the contrast vivid. Before antisepsis, surgeons were brave and fast, and patients often died. Joseph Lister's introduction of carbolic acid sprays and sterilization in the 1860s did not involve new knives or cleverer stitches. It changed the *informational environment* of the wound: the microbial signals and infections that were derailing healing. By altering conditions rather than anatomy, Lister unlocked the body's existing capacity to repair itself more reliably. In retrospect, antisepsis feels obvious. At the time it was a conceptual revolution: the problem was not only what the surgeon did with their hands, but what invisible agents did after the surgeon left.

Bioelectric medicine proposes another such revolution. The problem is not only that tissue is missing, but that the collective has the wrong internal signals about what to build next.

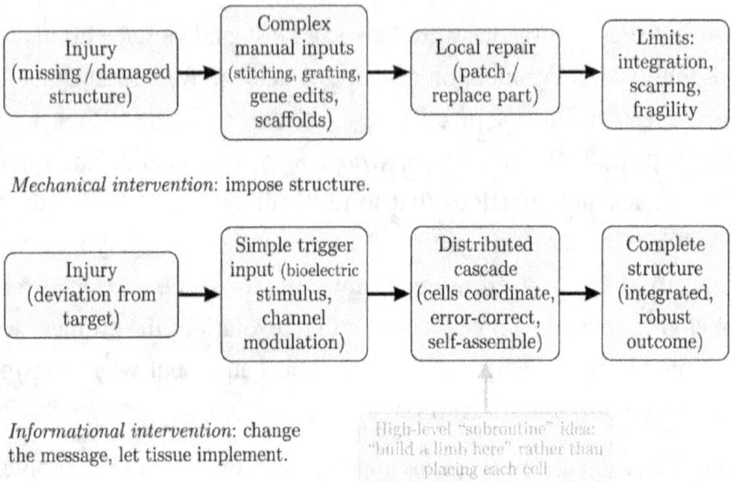

A sceptic might say: fine, but where is the actual "code"? In genetics, code is literal: sequences map to proteins. In bioelectricity, the mapping from voltage patterns to anatomical outcomes is not a simple dictionary. It is more like the mapping from a traffic-light pattern to the flow of a city. A single red light does not *contain* the whole plan of traffic. But patterns of signals can nonetheless steer collective behaviour toward stable large-scale structures. That is why the diplomat's medicine is as much about *interfaces* as it is about mechanisms. If we can discover which physiological patterns correspond to which high-level goals, we can begin to treat those patterns as a control language even before we fully understand every downstream molecular detail.

This is, of course, both thrilling and unsettling. Thrilling, because it suggests therapies that are more elegant than our current brute-force methods. Unsettling, because the moment you can call a subroutine, you can also call it in the wrong place, or call a novel one, or call it repeatedly. The body's construction crew is powerful. Power invites misuse, accidents, and ambiguity.

And ambiguity is already lurking in the background. If we can persuade tissues to build structures, we are no longer merely repairing damage; we are negotiating goals. Whose goals? The patient's, the physician's, the insurer's, the state's? What counts as enhancement rather than therapy? What obligations do we have when we create living structures that are meant to be partly autonomous, self-maintaining, and self-modifying?

Those questions sound like philosophy, but they arrive on the clinician's desk the moment programmable flesh becomes practical. The diplomat's medicine does not end with better wound healing. It ends with a new kind of responsibility: when you can rewrite the body's targets, you must decide which targets are permissible to write. That takes us directly to the next topic, where the power to program living matter forces ethics to grow new muscles.

9.2 The Ethics of Programmable Flesh

A few years ago, headlines announced that researchers had built "living robots" out of frog cells. The phrase did what phrases like that are designed to do: it made people picture something with a will and a mission, a tiny creature released into the world to do our bidding. The reality was both less cinematic and more philosophically awkward. Here were small clusters of living tissue, shaped by design choices and coaxed into motion by their own biology. They were not metal machines with batteries and code; they were made of cells that, in another context, would have become skin. Yet they also were not ordinary animals, because their form and function were selected by an external process. They were something in-between, and that in-between is where ethics begins to feel slippery.

The previous section treated a hopeful idea: that medicine might shift from forcing tissues into place to persuading them, with brief, informational nudges, to rebuild themselves. That hope comes with a quiet premise. If tissues can be guided toward a target form, then the target is not merely a passive outcome. It is a goal state the system is capable of pursuing, maintaining, and returning to when disturbed. In ordinary life we call that kind of behaviour *purposeful*. In the laboratory we reach for the cooler word *homeostasis*: the tendency of a living system to hold itself within a viable range. When homeostasis becomes anatomical—when tissue repairs a wound, or regenerates a limb, or corrects a patterning mistake—it looks uncomfortably like agency: not conscious choice, but a capacity to steer toward an outcome.

Once you accept that life comes with built-in goal-seeking at multiple scales, the moral landscape changes. Repair is no longer only a technical matter of tools and tissues. It becomes a question about what we are permitted to ask living matter to do.

There is a familiar version of that question in medicine already. A surgeon who amputates a gangrenous foot is altering the body for the patient's benefit, but also overriding what the body, left alone, would do. An oncologist who poisons a tumour with chemotherapy is deliberately inflicting harm in order to prevent a greater harm. We have ethical frameworks for this: informed consent, proportionality, oversight, and a commitment to minimize suffering. These frameworks assume a fairly clear boundary: there is a patient, who is a person, and there is a body, which is the patient's.

Programmable flesh blurs the boundary in two directions at once. First, it blurs the boundary between *treatment* and *redesign*. Second, it blurs the boundary between *organism* and *artifact*.

Start with redesign. If we can communicate with tissues in the language of pattern—bioelectric, biochemical, mechanical—then we can do more than fix what is broken. We can, in principle, alter the anatomical goals that tissues work toward. The difference between "regrow my fingertip" and "regrow my fingertip but longer" is not a difference of method; it is a difference of intent. In the mechanical era of medicine, enhancements were limited by what we could implant or attach. In the informational era, enhancement becomes a matter of changing the body's own self-assembly and self-repair routines.

The history of medicine contains cautionary echoes of what happens when a powerful capability spreads faster than the norms that guide it. The most painful example is the twentieth-century eugenics movement. Under the banner of "improvement," governments and institutions used medical authority to impose sterilizations and other harms, often on the poor, the disabled, and racial minorities. The lesson is not that all enhancement is evil. The lesson is that when biology becomes a site of optimization, the optimizer's values matter, and the optimizer is rarely neutral. A technique that can "improve" bodies will be pulled, inevitably, into existing social biases about which bodies are acceptable, employable, insurable, or worthy.

Programmable regeneration adds a new twist. Because the intervention may be indirect and distributed, it can also become opaque. If you replace a hip joint, you can point to the replacement and say what was done. If you nudge tissue into a new target morphology by altering network-level signals, the change is written into the body's living dynamics. It may be difficult to fully predict, difficult to fully reverse, and difficult to explain in simple terms to a patient. Ethical medicine depends on more than outcomes; it depends on the patient's understanding and voluntary agreement. But what does informed consent look like when the best honest description is, "We will send a signal that usually encourages the tissue to reorganize itself in the desired direction, although individual variation means we cannot specify every detail of what will happen"?

Some clinicians already face versions of this problem. Gene therapies can have long-term effects; immune-modulating drugs can reshape physiology for years; neural implants can alter experience in subtle ways. Programmable flesh pushes these issues into the centre. The more intimate and integrated the intervention, the harder it becomes to treat it as a discrete procedure and the easier it becomes to treat it as a permanent alteration of the patient's embodied self. That raises questions that sound abstract until you put them in a hospital room: how much uncertainty is acceptable? how much follow-up is owed? what counts as a side effect when the "device" is the tissue itself?

Now consider the second blur: organism versus artifact. When you build with steel, you are free to treat your materials as property. When you build with living cells, the moral intuitions that have guided laboratory work begin to wobble. In most biomedical research, cells in a dish are not treated as moral patients. We do not worry about their preferences. We do not wonder whether they are being wronged. We destroy them without ceremony. Yet the moment we start assembling cells into novel architectures that move, heal, and adapt, we begin to sense that the old categories are being stretched.

Part of the discomfort comes from an ancient philosophical knot often called the "problem of other minds." We cannot directly observe consciousness in another being; we infer it from behaviour, anatomy, and evolutionary continuity. That inference feels easy with mammals and harder with insects. With a cluster of cells designed in a lab, it feels almost absurd. Surely it is not conscious. And yet consciousness is not the only morally relevant property. Suffering matters, but so do interests, flourishing, and the capacity to be harmed by the frustration of goals. If we take seriously the idea that agency comes in degrees—that a tissue can pursue a target morphology the way a thermostat pursues a temperature, only vastly more richly—then the moral questions become questions about *which degrees* demand which forms of care.

A mechanical object has no stake in what happens to it. A dog does. Between them lies a broad grey zone: bacteria that move toward nutrients, plants that respond to damage, immune systems that remember, tissues that correct errors. Most of the time we ignore the moral significance of these goal-like behaviours because it would paralyse everyday life. We eat plants. We sterilize surfaces. We kill bacteria by the billions. Society runs on that moral simplification. Programmable flesh forces the simplification to be re-examined because we are no longer merely killing; we are *creating*, and we are creating entities precisely for their capacity to behave adaptively.

The question, then, is not "Are these new living constructs persons?" That question is too blunt and invites melodrama. The sharper question is: what obligations come with bringing into existence an entity that can act to preserve itself, seek certain outcomes, or respond to its environment in ways that resemble minimal preference?

There are at least two ways to go wrong here. One is to anthropomorphize: to project rich inner lives onto systems that almost certainly do not have them, and to let science policy be driven by horror and fantasy. The other is to dismiss: to assume that anything without a brain is morally inert, and to treat living matter as an ethically empty substrate for design, like silicone or plastic.

Ethics, at its best, is a technology for steering between those errors.

Concrete cases help. Consider organoids: small, lab-grown clusters of human cells that can develop features reminiscent of organs. Brain organoids, in particular, have raised questions about whether they might eventually exhibit neural activity patterns that deserve special scrutiny. Here the concern is not that an organoid is likely to have a mind like ours, but that the moral cost of being wrong could be high. If there is even a small chance of creating a system capable of unpleasant experiences, it is worth building safeguards early: limits on maturation, monitoring of activity, and clear oversight protocols. The pattern is familiar from animal research ethics: uncertainty does not remove responsibility; it increases the need for caution.

Now consider the engineered living constructs sometimes called xenobots. These were made from frog embryonic cells arranged into shapes that can move and, in some versions, push pellets or perform simple tasks in water. Their significance is not that they are clever. Their significance is that they are *designed living systems* that do not correspond to an evolved species and are not merely modified versions of an existing organism. That novelty breaks many inherited assumptions. Is such an entity an organism, a robot, a laboratory tool, or a petri-dish artefact? Different answers imply different oversight. "Organism" suggests ecological risk and animal welfare categories. "Robot" suggests product safety and cybersecurity metaphors. "Tool" suggests minimal moral concern. The words we choose become policy.

Even if you believe these constructs have no morally relevant experiences, there are still ethical issues that do not depend on inner life. Release into the environment, intentional or accidental, becomes a question. So does dual use: the possibility that techniques developed for healing could be repurposed for harm. Self-assembling or self-repairing living systems are attractive not only to regenerative medicine but also to militaries and industries. A technology that makes tissues robust against damage is, in principle, a technology that can make systems robust in hostile settings. Ethical governance has to anticipate these pressures without assuming that every innovation is a weapon in disguise.

There is also an ethics of *classification.* Modern regulation is built on categories: medical device, drug, biologic, tissue, organism. Programmable flesh refuses to sit still inside any one category. An electroactive scaffold that delivers patterned stimulation to guide regeneration: is it a device, a drug, or a combined product? A patch of engineered tissue that integrates into a patient and then continues to adapt: is it a transplant, a therapy, or a self-modifying implant? Categories matter because they determine what evidence is required, who is responsible when things go wrong, and how risks are communicated to patients.

A deeper ethical tension lies beneath these practical questions. It is the tension between treating bodies as *projects* and treating bodies as *selves.*

Medicine has always lived in that tension. Cosmetic surgery, fertility treatments, hormone therapies, prosthetics, and psychiatric drugs all raise questions about what kinds of bodily change count as health, what kinds count as preference, and whose standards are being applied. Programmable morphogenesis makes the tension sharper because it reaches into the body's own pattern-maintenance layer. Altering those patterning signals can change not only function but form, and potentially in ways that propagate over time. If a patient's tissues are guided into a new stable configuration, the change is not an external add-on. It becomes part of the patient's bodily identity.

This matters for consent in a particularly intimate way. A patient can consent to a prosthetic limb because it is a separate object with a known function and a clear boundary. Consenting to a self-organizing intervention is closer to consenting to a shift in the body's ongoing narrative. It demands not only a risk-benefit calculation but a discussion of values: what changes would feel acceptable? what degree of unpredictability is tolerable? what does the patient consider to be "themselves"?

Equity is the other unavoidable axis. Advanced regenerative therapies are expensive. If programmable flesh makes certain enhancements possible—stronger tendons, faster wound healing, resistance to degenerative disease—these will not spread evenly. They will cluster in wealthy populations first, as most technologies do. The ethical risk is not merely unfairness in access. It is the possibility of creating new social divisions that become biologically entrenched: groups whose bodies can literally repair and adapt better than others. That is a science-fiction scenario that becomes less fictional the moment we speak casually about upgrading biological goals.

At this point a public debate often pivots to a simpler question: "Is bioelectricity really a code?" The answer matters ethically because "code" implies readability, writability, and deliberate design. If there is a true morphogenetic code, then reprogramming anatomy becomes a plausible and powerful capability, and governance must treat it as such. If bioelectric interventions are better understood as one contributor among many—a way of nudging tissue dynamics without anything like a clear, general-purpose language of form—then the risks look more like those of other complex therapies: serious, but constrained.

The honest view is that both perspectives capture something. Bioelectric states clearly matter; they can gate downstream cascades and change outcomes. But the mapping from signal to shape is unlikely to be as neat as the genetic code, and it will probably remain context-dependent. Ethical prudence does not require certainty about the code metaphor. It requires a recognition that even partial programmability can have large consequences when applied to living systems that amplify small inputs into large-scale form.

There is a final moral problem that is easy to miss because it is not dramatic. It is the ethics of *responsibility for ongoing behaviour*. When you implant a pacemaker, you monitor it and replace it. When you seed a self-healing tissue construct that integrates into the patient and continues to change, who is responsible for its long-term trajectory? The clinician who applied the initial stimulus? The company that manufactured the scaffold or the ion-channel modulator? The patient, who now lives with a self-modifying biological system? Legal systems are not built for therapies that behave less like pills and more like partners.

Diplomacy again becomes a useful metaphor, but now with a darker edge. Diplomats do not only send messages; they also create obligations. Once you open a channel of negotiation, you cannot pretend that the other side is inert. If tissues and engineered living constructs are capable of adaptive, goal-directed behaviour, then creating them and altering their goals is not merely a technical act. It is an act that changes what kinds of relationships are possible between humans and living matter.

That realization can provoke either fear or humility. Fear says: shut it down, the categories are too unstable. Humility says: build better categories, better oversight, and better moral vocabulary. The second response is harder, but it is the one that fits the situation. Programmable flesh will not be governed well by panic, and it will not be governed well by complacency. It will be governed well only if we learn to describe degrees of agency, degrees of uncertainty, and degrees of obligation without collapsing everything into "just cells" or "basically animals."

One way to make progress is to shift attention from the noisy question of consciousness to the quieter question of *cognitive horizon*: how far a system can sense, predict, and act in space and time. A thermostat has a tiny horizon. A dog has a wider one. A human's extends to decades and continents. Tissues and engineered constructs sit somewhere between, and where they sit determines both what they can do and what risks they pose.

That is the bridge to the next section. Ethics needs a map, not of souls, but of capacities: what kinds of goals different systems can hold, how far those goals reach, and how those reaches expand when cells become tissues and tissues become organisms. Without that map we will argue endlessly about labels. With it, we can begin to see a continuum of intelligence that does not begin with brains, and does not end with them either.

9.3 The Cognitive Horizon

A cell does not have a brain. It has no neurons, no synapses, no internal narrator. And yet, in the previous pages, we have watched cells behave in ways that sound suspiciously like problem-solving: they coordinate with neighbours, correct errors, and converge on stable anatomical outcomes. If this sounds like an abuse of language, that discomfort is the point. The words we inherited for intelligence were forged in the presence of animals with faces, eyes, and obvious intentions. The moment we apply those words to tissues, the words either stretch, or they snap.

There is a way to stretch them without turning biology into metaphor. It begins by stripping intelligence down to something almost embarrassingly plain: *the ability to achieve goals in the face of uncertainty*. In this view, intelligence is not a substance you either possess or lack. It is a kind of control: the capacity of a system to act so that the world ends up in some preferred set of states rather than others. The preference may be simple ("avoid bursting") or rich ("write a symphony"). The mechanism may be fast (a reflex) or slow (a developmental program). What matters is the relationship between sensing and acting: can the system use information about its situation to steer outcomes?

Once you start looking for that relationship, brains cease to be the definition of intelligence and become one particularly powerful way of implementing it. This is the kernel of what some researchers call the scale-free or scale-invariant view of intelligence: the claim that goal-directed behaviour is not a special magic added at the moment neurons appear, but a property that can emerge wherever matter is organized into feedback loops that sense, decide, and act. Neurons are not the beginning of agency; they are an accelerator.

That shift can sound like wordplay until you apply it to something concrete. Consider wound healing. When skin is cut, cells at the edge do not merely divide randomly until the gap is filled. They migrate, they align, they change their adhesions, and they coordinate with immune cells and blood vessels. If the wound is small, it closes. If it is too large, the tissue may scar. If infection is present, the process changes. The system reacts differently to different conditions, aiming for a particular outcome: restore an intact barrier. That outcome is not guaranteed, and the fact that it sometimes fails does not make the process non-intelligent; it makes it a control problem under constraints.

Or consider a more dramatic example that has fascinated biologists for centuries: regeneration. When a salamander regrows a limb, the cells are not simply executing a rigid blueprint like a factory robot. They are responding to what is missing. They grow a structure of the right size and orientation, with nerves and vasculature integrated into the existing body. They stop when the limb is complete. These are the hallmarks of a goal-seeking system: responsiveness, integration, and termination.

The obvious objection is that we are smuggling intelligence in by describing outcomes as goals. Perhaps the cells are just obeying local chemical gradients, and the limb is an accidental consequence. But "local rules produce global order" is not the opposite of intelligence; it is the normal way intelligence works. Your immune system does not consult a central committee before mounting a response. Ant colonies build nests without a foreman. Even in brains, many behaviours emerge from local interactions among neurons. The question is not whether local rules exist, but whether the system's organisation allows it to *correct* deviations and *maintain* desired states.

A useful historical anecdote comes from cybernetics, the mid-twentieth-century science of control and communication. Norbert Wiener and his colleagues were trying to understand anti-aircraft systems that could predict the future position of a moving plane. Their key insight was feedback: the system must measure error (how far it is from the target) and adjust its actions to reduce that error. The mathematics that emerged was not specific to guns. It applied to thermostats, steering engines, and eventually to physiology. The body, they realized, is saturated with feedback loops, from blood sugar regulation to posture control.

But there is an even more unsettling implication of feedback: once you have it, you can build systems that appear purposeful without any inner experience. A thermostat pursues a temperature without feeling warmth. A self-driving car pursues a destination without longing for arrival. If tissues pursue anatomical completion, it does not mean tissues are conscious. It means that purpose-like behaviour can be a property of organised matter.

This brings us to the idea of a *cognitive horizon*. If intelligence is goal-directed control, then different systems differ not only in how much intelligence they have, but in *how far their intelligence reaches*. A creature with good senses and a fast nervous system can take into account events happening far away and anticipate consequences far in the future. A cell, by contrast, lives in a small neighbourhood. It can sense chemical concentrations, mechanical tension, electrical potentials across its membrane, and messages from adjacent cells. Its actions, too, are local: change shape, divide, move, secrete. From this perspective, the difference between a cell and a dog is not that one has intelligence and the other does not. It is that the dog's control loops span larger distances and longer times.

This is not merely a poetic way of talking. It gives you a handle on a question that otherwise becomes a shouting match: when does agency begin? The cognitive-horizon view lets you answer without drawing a sharp line. Agency begins wherever a system can sense a difference between "better" and "worse" states (for itself, or for the larger collective it belongs to) and can act to reduce that difference. What changes with scale is the breadth of what counts as relevant information and the repertoire of actions available.

The link to bioelectricity and morphogenesis becomes clearer if we think about what it means for cells to form a tissue. A single cell has a narrow horizon: it can keep itself alive, perhaps move toward food, perhaps avoid toxins. A tissue is a collective, and collectives can have horizons larger than those of their members. Not because a tissue is a mysterious super-mind, but because coupling lets information travel farther and lets actions be coordinated. Gap junctions and bioelectric domains are one physical way to achieve this coupling. They allow a disturbance in one region to influence decisions in another. In effect, they enlarge the neighbourhood that any one cell can "care about."

This is where the software metaphor becomes helpful again, if handled carefully. The software of a computer is not a ghost that makes circuits intelligent. It is a pattern of states that shapes what the hardware does next. In tissues, long-lived physiological states—including bioelectric patterns—can act like a memory that shapes what collective behaviours are favoured. If those states are stable, history-dependent, and switchable, then they can store something like a tissue-level context: not a sentence in English, but a bias toward one anatomical outcome rather than another.

Yet caution is essential. There is a tempting slogan in discussions of biology and computation: "the body computes." Like many slogans, it is true in a way that can mislead. If "compute" means "process information," then any feedback system computes. But if it means "carry out symbolic operations like a laptop," then most biological organisation does not fit. A limb does not multiply matrices; it regulates growth. A wound does not solve equations; it closes. The deeper point is that information processing in living systems is embodied: it is inseparable from the physical constraints and opportunities of tissue. The shape of the system is part of the way it controls itself. This is why some researchers prefer to speak of *morphological computation* only with careful qualifications: the body's form makes certain behaviours easy and others hard, but that is not identical to computation in the strict sense.

What the cognitive-horizon view offers is a middle path. It neither reduces intelligence to brains nor inflates every biological process into literal thought. It asks: what goals are being pursued, over what spatial and temporal scale, using what mechanisms of sensing and action?

A concrete modern example makes the stakes feel less abstract. Think about cancer. Cancer is often described as cells "going rogue," breaking the cooperative rules of multicellular life. That phrase can be sentimental, but it points to something real: multicellularity depends on cells giving up some individual goals in order to serve a tissue-level goal. A skin cell should not divide endlessly; it should divide when needed and stop when the tissue is complete. A liver cell should contribute to liver function, not maximize its own replication at the organism's expense.

On the cognitive-horizon view, cancer can be seen as a breakdown in scale coordination: the cell's effective horizon shrinks back toward unicellular priorities. The tissue's larger goals no longer constrain the cell's actions. This framing does not replace the molecular biology of mutations, signalling pathways, and immune evasion. It complements it by highlighting what is lost: not just normal gene regulation, but the multi-scale control that makes an organism an organism.

This is also why the bioelectric interface is medically tempting. If some aspects of tissue-scale control are implemented through physiological coupling, then restoring or modifying that coupling might help reassert the larger goal over the smaller one. The ethical questions of the previous section return here in a new form: if we can rewrite tissue-level goals, we must define what counts as a healthy goal. "Stop cancer" is easy to endorse. "Optimize body composition" is more ambiguous. "Increase tolerance to extreme environments" is, in the wrong hands, ominous. The power to tune the cognitive horizons of tissues—to expand or shrink what a collective can coordinate—is not value-neutral.

There is also a personal, almost existential consequence. Human exceptionalism often rests on the idea that minds are rare: that intelligence is an island surrounded by mindless matter. The cognitive-horizon view makes a different picture plausible. Minds like ours may be rare, but goal-directed control may be common wherever life organizes itself. In that picture, humans are not the sole bearers of agency but one dramatic point on a continuum.

Some readers will worry that this view cheapens intelligence. If everything is intelligent, nothing is. The worry is reasonable, and it highlights a linguistic trap. The goal is not to hand out gold stars to thermostats. The goal is to understand intelligence as a graded property, like speed or strength. A child, a dog, and a bacterium are all capable of control, but the scope and richness differ enormously. Calling them all intelligent does not deny the differences; it demands that we describe the differences precisely rather than using brains as the only gatekeeper.

A way to make that precision intuitive is to borrow a metaphor from physics: the light cone. In relativity, a light cone describes which events can influence, or be influenced by, a given event in space-time, given the finite speed of light. Replace the speed of light with the effective speed at which a system can sense and act, and you get an informal but powerful picture: each system has a cone of influence. A cell's cone is small: it can respond to local signals over short times. A dog's is larger: it can perceive distant objects, remember past events, anticipate near-future outcomes, and act across metres and minutes. A human's cone can extend across continents and decades, thanks to language, tools, and culture.

The point is not that biology obeys relativity in this sense, but that the cone picture forces you to ask two practical questions: how far can the system *know*, and how far can it *do*? When we say a system has a certain cognitive horizon, we are pointing to the size of its cone.

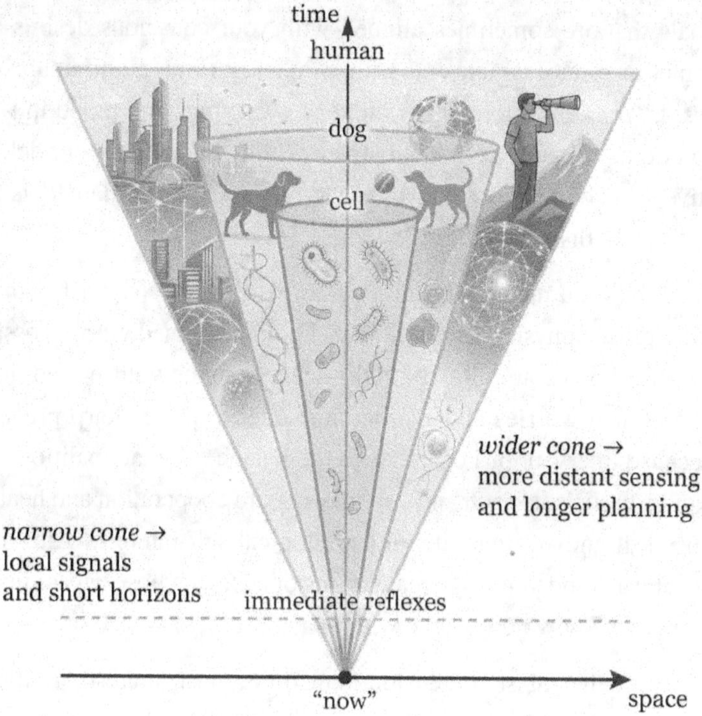

Seen this way, the story of life becomes a story of expanding cones. Single cells coordinate into tissues. Tissues coordinate into organs. Organs coordinate into organisms. Organisms coordinate into societies that store memory outside the body, in writing and technology, pushing the human cone outward in time and space. The expansion is not smooth or inevitable; it is achieved by new mechanisms of coupling and representation. Nerves are one mechanism. Hormones are another. Electrical coupling in tissues may be another. Culture is a mechanism too, albeit one built on the nervous system's scaffolding.

The scale-invariant view of intelligence has one more unsettling implication: if intelligence is about control over outcomes, then the boundary of "self" is also negotiable. Your immune system pursues goals that are sometimes aligned with your conscious desires and sometimes not (ask anyone with an allergy or an autoimmune disease). Your gut microbiome influences cravings and mood in ways you do not directly command. Even within your skull, your deliberative mind competes with habits and reflexes. The human "I" is less a monarch than a coalition.

That coalition metaphor returns us to where this chapter began: to the idea of a physician as diplomat. Diplomatic medicine is possible because bodies are already multi-agent systems with nested goals. The ethical worries about programmable flesh are sharp precisely because interventions can change the balance of that coalition. Expand a tissue's horizon and you may restore cooperation and healing. Shrink it and you may invite pathological self-interest. Alter it in novel ways and you may create forms of agency whose capacities we do not yet know how to live with.

For now, the safest conclusion is also the most interesting: intelligence is not a thing that suddenly appears when brains arrive. It is a graded capacity that grows as living matter learns to couple its parts into larger, more predictive, more far-sighted collectives. Brains are spectacular instruments for that coupling, but they are not the only ones.

The next question is unavoidable, and it will follow us beyond this chapter. If cones can expand, they can also be engineered to expand. When we build systems that are part organism, part artifact—whether a regenerative scaffold that negotiates with tissue or a designed living construct that persists in an environment—we are not only making new tools. We are shaping new cognitive horizons, and with them new kinds of responsibility.

www.ingramcontent.com/pod-product-compliance
Lightning Source LLC
LaVergne TN
LVHW022203200926
848481LV00006B/1832